A Life Lived
Remotely

A Life Lived Remotely

Being and Work in the Digital Age

Siobhan McKeown

Published by Repeater Books
An imprint of Watkins Media Ltd

19-21 Cecil Court
London
WC2N 4EZ
UK
www.repeaterbooks.com
A Repeater Books paperback original 2018
1

Distributed in the United States by Random House,
Inc., New York.

Cover design: Johnny Bull
Typography and typesetting: Josse Pickard
Typefaces: Neo Sans / Fedra Serif

ISBN: 9781910924785
Ebook ISBN: 9781910924792

Printed and bound in the United Kingdom

For Mark, who is missed

This is how it always ends,
With death.
But first there was life.
Hidden beneath the blah, blah, blah.
It is all settled beneath
the chitter chatter and the noise.
Silence and sentiment.
Emotion and fear.
The haggard, inconstant
flashes of beauty.
And then the wretched squalor
and miserable humanity.
All buried under the cover of the
embarrassment of being in the world.
Beyond there is what lies beyond.
I don't deal
with what lies beyond.
Therefore…
let this novel begin.

Jep Gambardella, *The Great Beauty*
(Dir. Paolo Sorrentino)

Contents

CRACKLES AND
DIAL TONES

I grew up feeling like everything had already happened. I was living at the end of history. It was too late, I had already missed out. As a teenager in the 1990s I can say, confidently, with hindsight, that it was a cultural deadzone. Nineties nostalgia makes me cringe: New Labour, the YBAS, and Britpop. The stand-out event was the rivalry between Blur and Oasis. I felt permanently disappointed – why couldn't I have been born earlier? Why couldn't my generation be defined by something more interesting? Everything felt so thoroughly boring, so settled. While the UK enjoyed a period of relative stability, I could only wonder "is this it?" I longed for the free-loving Sixties, punk of the Seventies, post-punk. Every band sounded like a poor repetition of earlier sounds. I was too young for the second summer of love and acid house; a week after I discovered Nirvana, Kurt Cobain shot himself in the head.

I was sure I had been born too late. I immersed myself in the music, art, and literature of the past, thought about gigs I would never go to, art movements that had already happened. But there was something new, though I wasn't able to pinpoint it then. It didn't come in the form of music or art, in protest or literature. It wasn't written or spoken or sung. It came along phone cables. It came in bytes. Its sound was crackles and dial tones. Its power was in connections.

In 1996, I was one of thirty-six million people who were connected to the internet. That's about 0.9% of the world population at the time. We gathered in a cobbled-together place, one of animated gifs and background wallpapers. Our websites were garish, built on services like Angelfire and Geocities. Our text was big and bold, words embellished with Comic Sans, Trebuchet, Webdings, Verdana, and Impact. Advertisements flashed and banners scrolled with glittering text. We were unafraid of background textures or looping animations, our headers were alight with pixelated flames. We used colour with the exuberance of a child with its first pot of finger paints, placed gifs like stickers everywhere possible. It was a time of great ugliness, but also of great possibility. There was a brave new world in front of us, and we few had access to it.

I entered the internet through AOL, the ubiquitous gateway that constructed its own singular version of the web. For me, the internet was all about chat. The sounds of AOL Instant Messenger (AIM) stay with me today: the creak of an opening door that signalled a friend arriving online, the *uh-oh* sound of an incoming message. I stayed up all night, light switched off so my parents didn't realise I was still up, hunched over the computer, talking to people all over the world.

It was incredible. I could speak to people I had never met, in places that I had never been. From my cold, dark room in the middle of nowhere outside Belfast, I was connected to the rest of the world, to the exclusive community of the internet.

In those nascent days of the World Wide Web, the internet was associated with pale-faced kids with big glasses who spent their lunch break in the computer lab. I kept my internet use to myself, maintaining a clear separation between IRL (in real life) and online. I had my teenage life where I would listen to The Doors and Pink Floyd and dream about bands so far before my time, and my online life, where I used Napster and Audiogalaxy to download their music. It was a magical time. I didn't come of age at an end; it was a beginning. The defining moment of my generation wasn't in culture or politics, it was in technology, and, for better or worse, it's had a greater impact on our world than most of the things that I longed for.

Twenty years since I first logged on, and ten since my first online job, I, like everyone else, am hyperconnected. I don't just log on anymore to chat to random people in other countries: my life is transformed by the internet. I am connected in every way possible, through my laptop, my phone, my tablet, my games console, my television, my watch. I almost exclusively shop online. I communicate online, via email, blogs, messenger, or services like FaceTime or Skype. I use the internet to figure out where to eat or go on holiday. I stand in the middle of unknown cities and use my phone to find the best restaurants within walking distance–without an internet connection in a new place I am lost. If I need to know something I look it up on Wikipedia. I stream music, watch films, read books and comics. I play womb noises on YouTube to sooth my crying child. I tweet on Twitter, blog on my blog, post semi-regular snippets of my life on Facebook. And, perhaps most importantly, I work.

It has been more than seven years since I worked in anything resembling an office. Like an ever-increasing number of people, my office is wherever I park my laptop. It can be my office at home, my sofa in front of the television, a coffee shop, a train station, a bar, an airplane, and sometimes even a beach. 95% of my communication is done using text, via chatrooms and messaging services. I work with people all over the world, in countries as disparate as Australia, Singapore, South Africa, Bulgaria, and the United States. Sometimes we meet, but most of the time we work from a distance, separated from each other but so connected that it's often easier and quicker for them to reach me than for someone located in my home town.

The world has shrunk, places that were once far away are now close. I can make friends in any country, in any time zone, and we can communicate synchronously or asynchronously. The world may be smaller, bringing together things that were once remote, but with our heads in the clouds it's easy for those things that are physically close to feel further away.

This book is formed from my reflections on working exclusively online since 2010. It is part-memoir, part-social theory, part-cultural theory, part-analysis. It reflects my own experiences of working online, and I hope that it resonates with the experience of others who work in similar conditions. It follows my own trajectory from when I quit my job to become an online freelancer, and uses my journey to explore the different issues at stake for people who work online. It is a book about work, about how the internet has changed the conditions for work, and the forces beyond work that have helped to prepare us for this world. Its main themes will be familiar to any remote worker: freelancing and remote companies, telework, relationships, communication, productivity, space, and travel. I have not stayed purely within the realm of the remote worker. To understand where we are today, to ask questions and interrogate these lives we lead, we need to look under the surface, beyond the everyday. A large part of this book looks at the technological and political changes that have created the conditions for the remote worker, and those invisible ideologies that create the fabric of the real. It is not a book that is intended to tell you how to work remotely, how to get a remote job, or how to run your online business. It is a book intended to help you reflect upon the conditions of work in this digital age.

Framed within the context of my own first steps into remote work, Part One explores the issues faced by freelancers, the realities of telework and spending all of your time at home, and how this can affect our relationships. Part Two deals with the technological and political conditions which have contributed to the shift to working online and the general casualisation of work. Part Three looks specifically at remote companies, issues around online communication, and digital nomads. There are many issues to do with work in the internet age which I don't cover. Topics such as the gig economy and automation are only touched upon because they are tangential to my own experience and have been covered extensively elsewhere.

It was important to me to write this book now because we are going through a transitional period, one in which concepts such as work and home are being redefined, and one in which we are being changed both by the technology we use daily and by the political and economic landscape we exist within. History has been marked by a series of such transitions, when technological, scientific, political, and religious changes have left people of that era feeling as though their worlds have been turned upside-down. I recall learning at school about technological changes and wondering what it would have been like to live through such a transition: how did it feel to hear that first radio broadcast, to travel on the first steam train, to read the first newspaper? What was it like to see the world drastically change before your eyes? The Victorians were particularly preoccupied with the fast and sweeping changes to their world. In a speech in 1860 to the British Medical Society the physician James Crichton Browne said "over the course of one month, our brains must process more information than our grandfathers had to in a few years of their lives"; while, in the 1890s, Clifford Allbutt attributed the belief in nervous illnesses to "the fretfulness, the melancholy, the unrest due to living at a high pressure, to the whirl of the railway, the pelting of telegrams, the strife of business, the hunger for riches, the lust of vulgar minds for coarse and instant pleasures, the decay of those controlling ethics handed down from statelier and more steadfast generations."

Now, it feels as though things are moving even faster still. Change is exponential. We live with similar concerns and similar preoccupations, when our thought needs to move at ever faster speeds as we are bombarded with more and more information. Just like the Victorians, we are living through a transition brought on by changes to the technological landscape that often make the world feel overwhelming. This transformation is happening quickly, exponentially, and we are changing with it. My generation is buffeted, forced to constantly adapt; it's important to pause and reflect upon where we have been, where we are, and where we are going, before we're something else. We millennials, much-maligned as we are, are the first, and only, generation to grow up with the birth and spread of the internet. For the generations of my children and their children, using the internet will be as unremarkable as picking up the telephone or watching television was to me. By then it won't be novel, it will just be infrastructure. We aren't there yet, but we are on our way: when the outsourcing of our lives to the internet is complete, I want to remember how we got there.

PART ONE
ENTERING THE ELECTRONIC COTTAGE

CHAPTER ONE
TOAD WORK

Most of us can look back at our lives and identify turning points. Not the big life events, like weddings and births and deaths. Everyday moments, unremarkable at the time but later on you think "yes, that's when it happened." I had one such moment in 2010. I had a job doing bookkeeping and admin at a restaurant in a remote corner of south-east England. We had moved a year previously when my husband, D, started a new job. My job was not glamorous. There wasn't a lot to do. I worked for the archetypal pub landlord; gruff, self-important, and brimming with dad jokes. I spent my days bookkeeping, writing convoluted drinks menus, patching together the website, writing blog posts, and staring out the panoramic windows at the white cliffs of Dover and the changing colours of the English Channel. I was scorched in summer, and freezing in winter, but mostly I was bored. I passed my time surfing the internet and daydreaming about what else I could do with my time.

The day it all changed is vivid, though it was like

hundreds of days before it. I was sat in my uncomfortable office chair, browsing crap on the internet. I stumbled across a website called Elance. Elance is a job-bidding site: people post their jobs and freelancers bid to win them. The list of jobs varies hugely, everything from web development, application development, design, writing articles, writing reviews, ghostwriting, copywriting, virtual assistant work, translation, email writing, technical writing, market research, typesetting, and so on. I recall my train of thought while browsing the advertised jobs: this is something I can do–there is nothing separating me from all of the other thousands of people already making money on the site. I spend most of my day faffing around in a dreary office, why not spend the time doing something else? Why not throw up a profile? What do I have to lose? As I filled in my profile, trying to think how I could stretch my previous writing experience (web content for previous employers, marketing bumpf, an article in an archaeology magazine) into a fully-fleshed out profile, I had no idea that it was the first step on a journey that would take my life online. In the space of a few years I would have friends and colleagues in countries I'd never been to, I'd have stumbled into the tech industry, and have adventures travelling the world.

Until that moment, working from home was a pipe dream. It wasn't for someone like me. How could I be someone who didn't have to deal with the inconveniences of work, someone who could roll out of bed at lunch, live in my pyjamas, hang out with my cats? I dealt with other people's business, doing work I had no stake in. As a responsible person who needed money to eat and pay rent, I had no choice. With my lack of monetisable skills what else could I do? My philosophy Master's may have left me able to talk around the intricacies of Spinoza's *Ethics* but no one wanted to pay me to do that. Work that involved going to an office, interacting with other people, selling my time, was necessary. I had to do it so I did it. I went in to my job, trudged down the hill in all weathers, sat in a room, and spent hours a day involved in tasks that were meaningless.

A few years previously, my plan had been to pursue an academic career. I was 100% invested. I constructed my identity around it. Towards the end of my degree I became disillusioned, and when I abandoned the academic life I lost not only my way but my sense of who I was. I didn't know what I wanted to do, who I wanted to be. I was unanchored–as dispersed and diffuse as the grey clouds that hung over the English Channel. Work was bound up with my identity. I had always wanted to *do* something; I wanted to *be* someone. Now, if someone were to ask what I did I didn't have an answer. I was a person, doing a job, like everyone else. I envied those who could say "I'm a doctor" or "I'm an architect" or "I'm a teacher," those whose roles and lives were clear and defined. I had no calling. Work was just a means to pay the bills. I went to work every day and, with Philip Larkin, would ask "Why should I let the toad work squat on my life?"

This tension between what I wanted–meaningful work–and the work that I had–toad work–reflects society's attitude towards work. Work as a concept, and the role that work plays in our lives, has evolved through history. In the Judeo-Christian tradition, work is a punishment given to Adam and Eve for sinning in the Garden of Eden–Adam is told that he will need to toil for everything that he eats. The Ancient Greeks also saw work as a curse. Aristotle, for example, saw the "mechanic or mercantile life" as being "ignoble and inimical to virtue." Without work, one could participate in the political domain and enjoy a contemplative life, while work takes time away from more noble pursuits.

This view of work as a curse continued for centuries, well into the medieval period. Labour served a church-supported established order and was condemned as a form of servitude. Confessional manuals, such as the *Summa Confessorum* of John of Freiburg, drew a correlation between what someone does and who they are. Every Christian is defined by their profession and each profession has its own list of sins; lust in innkeepers, gluttony in cooks, greed in merchants and lawyers. Such confessional manuals blend work with morality and identity, forming a key stage of the evolution that brought together what we do for money with who we are. Today we do not equate work with sin but the correlation between work and identity remains. It's no coincidence that around the twelfth century surnames started to be used that identified people with their work–Baker, Carpenter, Cook, etc.

From the idea of work and sin emerged the idea of work and salvation. A profession isn't just fraught with the dangers of sin, it is a path to salvation. Through work you can save your soul. In the sixteenth century Martin Luther and the Protestant movement redefined work as something you do for the benefit of yourself and of society. A founding principle of the Protestant work ethic, which underpins our modern conception of work, is that work is something good, something that you define yourself by, it demonstrates your integrity, your goodness, your responsibility, your position in society.

Today, work is often the means through which we construct our identity. We spend more time doing work than we spend doing any other single thing. Many aspects of work help us to construct a stable identity, providing the conditions for us to act in the world. This includes where we work, what we do, the markers of a specific job, whether that be a uniform or the culture of a workplace. We produce work, but in doing so we produce our selves. If our work is demeaning or unfulfilling we find not just our work but our lives diminished. We are known by what we do, we introduce our friends and peers to one another by describing our jobs.

What a person does for money, however, does not necessarily reflect who they are, and the way that society entwines work and identity conceals the complexity of human identity and experience. Perhaps during the day you do data entry but in the evening you do stand-up comedy, you work behind a bar but write poetry, you are a cleaner but you are an accomplished painter. By which should you be defined: by the thing that enables you to eat day-to-day or the thing that you care most about?

I recall being asked as a child: "what do you want to be when you grow up?" Most children respond with something straightforward: a vet, a doctor, a fireman, an astronaut, an actor, a clown, a racing-car driver, a teacher. The question seems innocent, but asking it sets up an identity between your job and who you are. Growing up, the answer to the question might change – the vet becomes a zookeeper becomes a nurse becomes a lawyer – but the promise of the question does not: that you will grow up and have a fixed job that provides you with a stable identity for the rest of your life.

A more accurate answer to the question might be: "I'd like to spend a few years working at Tesco, then get a job doing admin in an office, after that I'll train to become a fitness instructor." We start our lives with an expectation that work will fix our identity but work is complex and what we do changes through different periods of our lives. Furthermore, the conditions of work and our ability to work often originate outside us, which makes it an unstable foundation for an identity. If a person loses their job, whatever the reason, it can precipitate a period of self-questioning and depression, a loss of identity that can be difficult to recover from.

It's this type of position that I found myself in when I diverged from my original, fixed career path. My identity was thin and fragmented and I was unable to find or sustain work that I found meaningful. I could not grab onto that thing through which society promised my sense of self. And so the work I was left with was whatever I could do to earn enough money to buy food, get some nice things, and pass through life.

After more than a few years of drifting from admin job to admin job, I was sitting in that grey and shabby office and I spied, through Elance, a way out. Here's how Elance works: company posts a job, say twenty articles to be written on the topic of cheese. Freelancers respond with a cover letter outlining their price and why they should get the job. Company sifts through the cover letters, chooses the best fit for the job, and awards the job to a freelancer. Freelancer does work, gets paid, everyone is happy. Safeguards are in place for both the freelancer and the client. The funds for the job can be put into escrow (a third party that later disburses the money), ensuring that the freelancer is paid for their work. For hourly work there is a time tracker which emails reports to clients. It has a stalker-ish feature called Work View which takes screenshots of what the freelancer is doing and streams them into a shared virtual office. An incentive for using Work View is that payments are guaranteed; the freelancer is paid and the client can make sure their work gets done.

A company will find many advantages to using remote freelancers. The biggest is that freelancers are expendable. There is no obligation beyond the current contract. A company is able to sidestep all of the associated costs of having an employee: office and equipment costs, admin and accounting overheads, maternity pay, sick pay, training, and so on. There is a direct monetary transaction with no long-term commitment. Once the contract is over, the work completed, the relationship is over. Employees have rights that make them hard to get rid of. By outsourcing a certain amount of work to freelancers and contractors, a company has an easy way to shed employees if times get tough. A company with 90% employees and 10% contractors can instantly shed 10% of its workforce if necessary. The risks associated with work and employment are therefore shifted from the company to the workforce. In extreme cases, for example companies like Uber and Task Rabbit, the entire cost and risk associated with a worker is literally outsourced onto that individual themselves.

Beyond straightforward considerations around employment law there are other benefits to a company. Perhaps the company needs to augment its talent pool. A team may lack one specific skill or have an unusually high workload for a short period of time. A web-development company, for example, needs to hire a ux specialist for one project, or a copywriter to produce website content. A product team may hire in a marketer to deal with a product launch. There may not be space in the company for a full-time role but the company will return again and again to the same freelancers. Thus freelancers, through repeated work, can build up an ongoing and relatively stable income.

A freelancer comes to a company as a complete package, already an expert in their field. He or she is expected to work on a job with no, or minimal, training. The freelancers who command the highest rates are those with the most experience, who can instantly step into a role, bringing all of that experience with them. They bring a fresh set of eyes and a different perspective to a team that may have become stale. There's no hiring process, onboarding process, or training. Hiring a freelancer means getting someone in to get work done.

Just as there are benefits to companies, there are benefits to the freelancers. For one, there's a sense of empowerment that comes from working for yourself. In my early days as a freelancer I felt an overwhelming sense of freedom. It felt like the first step to the world promised by Marx: the internet made it possible for me to "do one thing today and another tomorrow, to hunt in the morning, fish in the afternoon, rear cattle in the evening, criticise after dinner, just as I have a mind, without ever becoming hunter, fisherman, herdsman or critic." I worked my own hours, in my own time. It gave me a sense of control over my life that before had been lacking. I didn't have to go and sit in some office, selling my time for someone else's ends. I did work, work that I chose, when and where I wanted to do it. And if I didn't want to work I didn't have to. Of course, the longer I freelanced the more I realised that I was subject to client timeframes and delays. There is a cost to the freedom of being a freelancer: juggling multiple jobs, intense task-switching, admin overhead, loss of security.

When the going is good, the loss of security doesn't matter. A good freelancer can command rates much higher than their employee equivalent, which works out well for both the freelancer and the client: for the freelancer because they're earning more than their employed counterparts, and for the company because they get flexible labour. Over time I discovered that the burden of risk and overhead had shifted from the company to me. I took on numerous financial and work burdens an employee doesn't have: I paid for equipment such as a laptop, tablets, and office furniture, software, training, and accounting services. I had extra admin overhead. I had to spend time finding new clients. I saw an increase in my utility bills. I no longer had any work-related benefits. I had no sick pay so I worked through illness. I didn't take holiday because time not working meant a big gap in my income. It didn't take long for me to realise that my high hourly rate was barely minimum wage. Any increase in my monthly income was due to the numbers of hours I worked, not because I was paid well.

For my first gigs on Elance, I worked both for companies and for individuals who needed help getting a specific project off the ground, one-person shows like me who were launching a website or product. My first ever job was for a web developer who had set up an online service for people to archive their family photos. He took a chance on me despite my lack of profile, put up my rates when he was happy with my work, and we worked together on and off over the following year. This initial good experience encouraged me to keep going. Not all of my gigs were so well-paid.

Because freelancers on Elance bid for gigs, people bid low for jobs. One of my first gigs was writing 24 articles of 280 words about restaurants in the UK. Each article focused on an individual restaurant, and since I hadn't been to most of them, I had to research each article. (Yes, in case you weren't aware, lots of the review content on the internet is written by people who haven't used a product or been to a restaurant or read a book, or whatever.) For this I was paid $200. It is what I had bid, so the pay was my own responsibility, but by the time I was 12 articles in I cursed myself for bidding so low and wasting my time.

Like so many internet platforms, a cornerstone of Elance is its review system. From eBay to Amazon to Airbnb to TripAdvisor, reviews are an accepted way for a person or business to establish trust. Whatever platform you are on, your profile starts with no reviews and you need to build it up over time. Elance's review system operates much like others, with clients leaving feedback on a freelancer's profile. The more reviews you have, the more your profile grows, and the more you can increase your rates. At the start, you bid low to get ratings so that you increase your future earning power. Freelancers can also leave reviews for clients, providing warnings about any who are difficult or who don't pay their bills.

Whether you gig on a platform like Elance, or pursue more traditional freelance work, the reputation system – a close network of people who recommend talent – is a mainstay of freelance life. A freelancer must maintain their reputation in this system, usually through the quality of their work. On a platform like Elance this means getting good reviews, but elsewhere previous clients may act as referees, provide website testimonials, or recommend you directly to friends and colleagues. The reputation system helps good freelancers find work but, since you are reliant on your reputation for your income, you must keep tight control of it. It doesn't matter how badly a client has treated you, if they badmouth you to others it can have a negative effect on your ability to earn.

Later in my freelance career, when I was no longer reliant on Elance but had built up a network of clients, I experienced the downsides of the reputation system. I had a solid reputation which meant I had lots of work and high rates. I was hired by a client to create some technical videos explaining their product. I had worked with them before and it had gone well so I didn't charge them a deposit. I laid out my terms, telling them I needed timely feedback in order to produce the work in the timeframe they required. I needed the work completed because I had another big project due to start in a few weeks' time. I recommended that we hire a voiceover artist: my own Northern Irish accent is never easy to understand for an American audience. They liked my accent so wanted me to do it. I produced a first draft of the videos, about ten hours of work, and I waited for their feedback. And I waited and waited. I sent follow-up emails and follow-up emails. I had turned down other clients so that I could get the work for this client done, but their lack of response meant I was doing nothing, earning no money. In the end, when I finally got feedback, it was that I had missed some important points and that they didn't understand my accent. They were unhappy with my work, and when I submitted my invoice, which was only for $560, they were only prepared to pay half. I felt angry, but I also felt powerless. I wasn't in a position to legally pursue them and, even if I was, it wouldn't be worth it for a couple of hundred dollars. I wanted to write on my blog about it, to talk about what a terrible client they had been and how they had treated me badly, but I knew that it would reflect poorly on me, giving other potential clients second thoughts about whether to work with me. In the end I did nothing: I wrote the entire invoice off, refused to work with the client again, and always charged a deposit, even for clients I knew well. For me, the incident cost just a few hundred dollars, some time, and some pride. But other freelancers find themselves in much more serious situations, waiting for payment of thousands, or even tens of thousands, of dollars. As a self-employed person you have few rights beyond what you've written into your contract, and if you haven't got a solid contract, or aren't able to pay for a lawyer, there's little that you can do when a client doesn't pay you. This is further complicated by working with global clients who may have different laws and regulations when it comes to employing freelancers. Calling out a client on the internet is generally seen as bad business etiquette and can have a negative impact on the freelancer's reputation. This attitude disempowers freelancers even further. Self-preservation in a crowded market often means staying quiet, even if it leaves you out-of-pocket.

A 2015 study of freelance media workers looked at other problems with the reputation system. Some complain that the reputation network is closed and doesn't allow for free competition for jobs. Others complain that it's too open – it's easy to be undercut by rivals who will accept less pay and poorer conditions. The reputation system causes anxiety amongst freelancers – how they are perceived in public life can have a real impact on their financial stability. Certainly when I was freelancing I found myself more cautious about what I said on public forums like blog comments and social media. This type of self-censorship has you second-guessing yourself. The most important thing, more than money or how you are treated, is the preservation of your reputation. This undermines the notion that a defining feature of being a freelancer is choice. You feel like you have choice until the moment you realise you don't, when a client screws you over and you have to keep your mouth shut to protect your reputation.

Using freelancers may help companies save money but there are broader repercussions that aren't always easy to predict. A troubling example is in the area of investigative journalism. A survey by Project

Word found that due to challenging conditions and inadequate support, freelance journalists are forced to abandon stories. The report found that 84% of respondents subsidise their work with their own finances, that 55% fail to recoup the costs of writing a story, and that 14% receive no compensation at all for their work. Freelance journalists aren't making enough money to cover their costs, and publications are less willing to cover expenses, paying just for the story itself. With freelancers paying out of their own pockets to investigate stories, many stories are abandoned when the journalist is unable to make ends meet. One former newspaper staffer, now freelancer, talks of surviving on "social security, food stamps, Medicaid." According to Project Word, many stories that are lost are in the public interest:

> Based on comments and interviews with respondents, such aborted stories covered a range of topics, from how the Pentagon handles the health of soldiers and the US role in mass killings in an ally country, to global reproductive rights, an investor's plan to bundle and sell minority-owned broadcaster companies on the public spectrum, and 'blatant corruption' in Florida.

Freelancing might be good for the business bottom line, but it isn't necessarily good for producing high-quality content. Stories are lost, important stories that provide the public with a deeper insight into what's going on in the world, stories that are desperately needed in a world abundant with trivial content. My own experience of freelance writing isn't much different. The in-depth, thoughtful pieces, the ones that take a long time to write, the ones that are well-researched and that I agonise over, are those least paid. Some of my best writing for online magazines has gone unpaid, or paid at a rate of £40 per article which, for the time I spend writing it, might as well be unpaid. Instead it's the listicles and the content marketing, the SEO articles, the churn of content, which is cost-effective to produce. I can knock something together in an hour from a bunch of dubious internet sources and get paid £40 for it, or pour in my heart and twenty hours and get paid the same: only one of these approaches is sustainable for a person who has to eat, pay rent, perhaps pay healthcare, perhaps look after a family.

These problems and reflections would come to me later, in a life beyond Elance. In those first months, and even years, of freelancing I felt free. I had a sense of satisfaction, a sense of control, and a sense of power. I browsed for the jobs that I was interested in, bid an amount that I was comfortable with. As those first few jobs trickled in, I felt the balance of power shift. I didn't have to grind away day by day, putting my time and labour to use for someone else. I could do it for myself. I would squish that fat toad. The more jobs I had, the more reviews I got, and the more jobs I won. My ability to get work snowballed; I went from doing a few paid gigs on the side, while I was bored at work, to taking on more, working weekends, working evenings. I was happy to do so because I wasn't doing it for anyone else: never mind that I was writing second-rate articles about web design or sales copy for iPhone applications, I was doing it for me.

CHAPTER TWO
THE PURITAN INSIDE ME

Christmas time. I took two weeks off to eat, lounge, read, and do bits and pieces of freelance work wearing my pyjamas on the sofa. As the end of these two weeks approached I was filled with dread. I couldn't bear the thought of going back to that dreary office, to look out at the chalky cliffs and the stony grey sea, to waste my time working in some mediocre restaurant. The last few days before going back I was miserable. I didn't yet feel confident enough to give up the stable income, and yet I was getting freelance work that paid me more per hour than I got at my job.

"You should just quit," said D, as 2010 turned into 2011 and the return to work, grey and gloomy, steadily approached. I'd had a glimpse of what was possible if I could just take the next step and quit my job: halcyon days, laptop on knee, pattering away at my keyboard with *Homes Under the Hammer* on in the background. Or I could go back to a job that did nothing but provide me with a wage. Every day that passed I felt more ready to pack it in. I was scared by the lack of security but I knew I needed to take a risk. It felt right. And so I quit. In doing so I put off the grim inevitability of being back in the office, of sitting at that desk staring at my computer, of seeing my boss. I put it off indefinitely.

My job was quickly a thing of the past, bringing to an end the five-year-long string of admin jobs that I'd been floating between since university. So many things that I'd hated about working were, in the space of that one decision, gone. No more going into an office. No more commute. No more sitting at a desk making up shit to do. No more work emails in the evening. No more committees. No more meetings. No more taking minutes. No more boss.

Freedom from managers was one of the most immediate benefits to giving up the normal nine-to-five. I have had a lot of bosses. I have disliked about 90% of them. Maybe I'm unlucky, or maybe there's something wrong with me, or maybe it's just the nature of the manager/managee relationship that turns a person into a posturing asshole.

When I worked at a call centre, selling household appliance insurance for a well-known gas and electric supplier, my team leader was a former us Marine. I have no idea what he was doing working in a uk call centre. He marched amongst us, stocky and bald-headed, as we sat heads-down at telephones that autodialled customer after customer. I recall him barking at us for slacking, his militaristic instruction in telephone sales, the way he championed the person with the best sales and yelled at the people with the least.

I've had many different types of bosses: officious bosses, those who are pedantic about time-keeping and process. Bosses who think that they are funny, who tell jokes that I'm forced to laugh at while inside I think "I fucking hate you." Bosses who are "just like you," who want to be my friend, one of the team, one of the guys, but whose facade disappears during staff appraisals or budget cuts. Patronising bosses who waste time by explaining every little detail, as though I am just a child who needs my hand held through my job. Bosses whose only input is to correct my spelling. Arrogant bosses who love to talk about themselves. Bosses who always know best. Bosses who are easily deflected with a winsome smile, bosses who are a bit creepy, angry bosses, stupid bosses, psychopathic bosses, bosses who are all of the above.

When I quit my job and became a full-time freelancer, I no longer had a boss. Bosses were a thing of the past. From now on I would be free and easy. No one to tell me what to do. No one looking over my shoulder, no one tallying up my work. No more grinding away, counting the minutes and hours before home time. I was in charge.

My first day at home with no job to go to. The glow of the rising sun seeps around the edge of the pale wooden blinds, bathing my room in a golden light. Beside me, D snores. It's the gentle, raspy snore he has in the morning, softer than the cavernous choking noises he makes throughout the night. On the floor beside his side of the bed, clothes are neatly folded and his iPad is plugged in, balanced against the wall. On my side, clothes are strewn everywhere; the half-worn mixed with the dirty, mixed with a pile of clean that I kicked over absentmindedly on my way to bed. My iPhone is plugged in, poking out the top of a half-inside-out pair of navy trousers. I reach for it. It sits snug in my hand. I lie on my side, still with the duvet pulled up to my ears and, with the brush of a finger, with a tightening of expectation, I connect. The phone is cupped closely to my face, as I tap tap tap at its screen, my back to D, my face to the internet.

As his snores deepen I skim through my email, eyes flickering over the bold text of my unopened messages. Spam and work. I open Facebook. A little red notification symbol tells me something has happened. There have been thirteen Likes of a photo I posted of D cradling our cat like it's a baby. I had given it the title "Madonna and Child," which I think is cute. I tap to find out who liked it, scan the names–the usual suspects–take a beat to register their approval, and then scan my news feed, mechanically tapping "Like" on anything that elicits more than a half-smile. In a few days I won't be able to tell you what any of those people are doing.

Facebook done, I open Twitter. No notifications, which is to be expected as I barely tweet. Still, I spend as much time on it as the most prolific tweeters. I scroll the thousands of tweets that have been posted since I last looked. Sometimes, rarely, there is an article that I'm interested in. I click it and save it to my endless backlog of reading on Pocket. I spend the next thirty minutes scanning tweets, hundreds of lines, barely registering the content of what people have said, forgetting who has said what the moment it has passed by in my stream.

D stirs.

"Morning," I say, as my finger continues to skim the

surface of the screen.

He rolls over. Grunts.

"What time is it?"

"7:30."

"Ugh." He smacks his lips together a few times, reaches out of the bed, grabs his iPad. "Too early," he says as he opens his email. We both look at our screens for another fifteen minutes before he groans again and gets out of bed. I tuck my phone under my pillow, snuggle deeper under the cover as he struggles on with his trousers, pulls a t-shirt over his head. Sucker.

"Turn the light off," I say, barely able to contain my grin, and I close my eyes once more.

There is a sense of quiet. A lack of urgency. I slip back in to a delicious stolen sleep, a precious extra hour in bed. I wake later and laze in a hypnopompic state, swimming through half-formed dreams.

When I finally get up, I pull on what will become my standard workday uniform–a pair of worn navy tracksuit bottoms, and an old baggy t-shirt. Over the years this uniform will have a few variations–yoga pants, pyjama pants, tracksuit bottoms, from band t-shirts to the increasing number of conference t-shirts that clutter my drawers and are the victims of frequent clear-outs to the charity shop. I go to the kitchen, make coffee, and slope back upstairs to my small office.

It is lined on two walls by Ikea Billy bookshelves. Cheap and functional, they hold the hundreds of books that D and I have accumulated throughout our years as voracious readers. They are filled with books that I'll never read, that I've come across in a second-hand bookshop and thought that I might like to read sometime, or that people have bought me as gifts and

I've never got around to reading. I am comforted being surrounded by all of these books. They have a sense of depth and solidity that contrasts with the ephemera of the internet. It's fashionable to proclaim the death of publishing, the end of the book in the face of the blog, but I'm not so pessimistic about the book's future. The book and the internet are different media, have a different form.

Books require a different type of engagement from both the reader and the writer. Each book is not just a collection of words but represents an extended period of time that the author has dedicated to thinking about their subject, conducting research, crafting prose. It requires focus to write and a level of determination to see it through to completion. A blog post, or internet article, is different: I have written many such articles and even the most in-depth didn't take more than a few weeks. A blog post can be dashed off; it is by its nature ephemeral, it has a time stamp, is part of a timeline.

Both are different in their consumption. There is a one-to-one relationship between the book and reader. When you read a book, you enter into its world on its terms, allowing yourself to be immersed. You may talk to people about the book afterwards, but the act of reading a book is a moment of solitude. Internet articles, whether they be posted on blogs or appear in online magazines, are always in the context of a broader conversation. The reader of the internet article has the opportunity to comment on it, whether that be in comment threads or by writing about it in social media. The internet is not just a medium for reading, but also for communication and other forms of media consumption. This means that there is much more potential to be distracted. The advent of the internet has made it possible for anyone to participate in conversations and to share their ideas, but this is a wholly different type of object than the book.

I calculate that if I read forty books in a year and live for another forty years I'll read sixteen hundred books. But forty books a year is ambitious. Maybe I'll read twenty, so eight hundred books. A depressingly finite number of books that I will read in my lifetime. That's not all I'll read though – I'll read many millions upon millions of words on the internet, tweets and Facebook posts, blog posts and comments, articles about things that I can barely remember. Some of this will be thoughtful and insightful and worth reading, but much of it will just be the ever-escalating, boundless heap of crap that is the internet.

I drop into my cat-clawed pleather chair and fire up my PC. From my window I watch my black fluffy cat stalk a frog in the garden. I settle down to work, a list of things I need to do scribbled down on a piece of paper. My first ongoing freelance gig was for a company in Australia, blogging about the WordPress content management system. I could do whatever hours I wanted, I just needed to produce articles and bill for my time every month.

I liked being paid for an object that I created. Most of the jobs I'd had previously involved administration, sales, manual labour, or service. Now I was actually creating something that had a value to someone. It didn't matter that I didn't know much about WordPress. Nor that I was selling myself short in terms of money and time. I felt in control and I was happy.

Since I was paid by the hour and I could work as

many hours as I wanted, I started to think about each hour of my time in terms of its monetary value. A simple calculation: should I spend a few hours watching television or should I use those few hours to make money? The first option was always a waste of time – why would I sit around doing nothing when I could be getting more cash? I had nothing to do at the weekend so I might as well work all day. Evenings were just dead time, passing hours until going to bed, might as well be making some money. I began to live the words of Benjamin Franklin: "Remember, that *time is money*. He that can earn ten shillings a day by his labor, and goes abroad, or sits idle, one half of that day, though he spends but sixpence during his diversion or idleness, ought not to reckon *that* the only expense; he has really spent, or rather thrown away, five shillings besides." I would not sit idly by, throwing away money.

I found myself spending more and more time at my computer, whether it was cloistered in my office or sat in front of the television. I was inundated with work, revelling in my new-found freedom, counting the money as it came in, every penny gained through my own hard work. My workload became colossal, and soon I found myself struggling with two of the curses that plague both freelancers and remote workers: multitasking and procrastination.

There is no better medium for multitasking than the World Wide Web. As communication media have changed and developed, they have required less and less focus and allowed us to do more while we are in the process of communicating. When handwriting a letter, I am alone with my pen and paper: there is a singular commitment to the task which forces me to focus. I may face distractions and interruptions, but that possibility is not contained within the medium itself. With the telephone there is a greater opportunity for multitasking: I have a telephone to my ear and am semi-engaged in the conversation but, because the person cannot see me, I can also be doodling on a notepad or cooking my dinner. The telephone creates a greater and more immediate sense of connection but, because it only concerns itself with our auditory senses, it leaves scope for doing other things, albeit in a limited capacity.

The move from speech to text-based communication creates limitless possibilities for multitasking. When engaged in a conversation, I can be doing many different things: reading tweets or Facebook statuses, responding to emails, watching a movie, talking to countless other people, reading Wikipedia. The medium itself does not demand my full attention.

The word "multitasking" has two definitions, one of which relates to people. To multitask is to deal with more than one task at the same time. The first published use of the term in writing, however, relates to computing. In computing, to multitask is when a computer performs multiple tasks at the same time. Computers are pretty great at this. They can run multiple different programmes, and increasingly so. The more powerful a computer is, the more tasks it can run at the same time. And computers, unlike human brains, keep getting more and more powerful.

On the internet, data is transferred along a packet-switched network. This means that data is broken down into chunks and transmitted simultaneously.

Internet data is fragmentary, broken down, and then reassembled when it arrives at a client computer. The mind of a multitasker starts to mirror the way the internet works. When I spend time on the internet, whether it's communicating, browsing, playing games, or working, I have a tendency to operate in this fragmentary way, picking up small pieces of information and processing them, hopping from half-thought to half-thought, half-finished task to half-finished task.

We multitask, approaching multiple tasks at once in an effort to do more or, in many cases, appear to be doing more. In a normal working day I have lots of chat clients open, giving people different ways to disturb me and ask me questions. Often these questions take thirty seconds to answer, but the constant possibility of interruption means that I can never really focus deeply on what I need to do.

The internet has increased the possibility for multitasking but, unlike computers, more power has not been added to our brain in order to cope with processing this number of tasks. To deal with this high number of tasks, we become reliant on external tools that augment our minds and help us to process large amounts of data. But because of the fast-changing nature of the internet, these tools are only useful for so long before they become defunct and we have to find new ones. And any tool that helps us to multitask only obfuscates the real problem: that multitasking doesn't help us be more productive at all.

As the internet has become a more permanent feature in our lives, so the research on multitasking has increased. More and more articles are published about the negative effects of multitasking. When you jump from task to task you feel a sense of accomplishment, of getting things done. I can easily spend a day responding to email or incoming messages but I never seem to get to the task that requires at least a few hours of unbroken, deep concentration.

While a computer can multitask, doing multiple tasks at the same time, it's mostly impossible for a human being: instead we are task-switching – switching from one task to another. Task-switching results in a problem called attention residue. When I switch tasks, part of my mind is still focused on the previous task. Every time I refocus I have to remind myself what I was doing. There has been lots of research into the negative effects of task-switching: it takes more time to get things done if I am constantly switching, the possibility of error increases and these errors increase the more complex the task. It can cause up to a 40% lose in productivity. A 2005 study at the Institute of Psychiatry found that people who are distracted by multiple tasks have a ten-point drop in their IQ; that's more than twice the impact of smoking marijuana. Another study of multitasking did MRI scans of the brains of multitaskers and found a decrease in brain density in the areas controlling empathy and emotion.

Our brains are hardwired to respond to novelty: whenever we respond to something new we get a hit of dopamine, the neurotransmitter which rewards us for taking action. It not only makes us feel good but keeps us searching for more novelty. This makes multitasking hard to resist. Our virtual environments provide us with perpetual novelty so we can constantly move to a new task. Our brains reward us for losing focus and for constantly moving on to the next shiny thing. Daniel J. Levitan writes in the *Guardian*:

> The irony here for those of us who are trying to focus amid competing activities is clear: the very brain region we need to rely on for staying on task is easily distracted. We answer the phone, look up something on the internet, check our email, send an SMS, and each of these things tweaks the novelty-seeking, reward-seeking centres of the brain, causing a burst of endogenous opioids (no wonder it feels so good!), all to the detriment of our staying on task. It is the ultimate empty-caloried brain candy. Instead of reaping the big rewards that come from sustained, focused effort, we instead reap empty rewards from completing a thousand little sugar-coated tasks.

Alongside this, multitasking increases production of the fight-or-flight hormone adrenaline, and cortisol, the stress hormone. The brain of the multitasker is searching constantly for the pleasure of novelty, but always anxious, always stressed. I can bring to mind many people I've met through working online who, unable to pause for even a moment, are always squirrelishly looking at their phones or repetitively opening and closing the lids of their laptops, seeking out their next hit.

As well juggling multiple different tasks our brains are constantly processing those tasks: which is the most important? What should I do first? My brain, at least, isn't always well-equipped to do this. I'll be halfway through writing a Facebook status, then look something up on Wikipedia, read half an article then check my email, half compose an email response, check Instagram on my phone, and click on the Facebook tab in my browser to realise that I was writing something there. Multitasking increases our cognitive load. As well as completing my tasks I have to process them at high volume. As well as processing and prioritising all of these tasks, I need to make decisions: should I respond to this email? Do I like this picture enough to click Like?

Despite evidence to the contrary, there are still people who insist that multitasking makes them more productive. To them, Earl Miller, a neuroscientist at MIT and an expert in multitasking, says "People who think they are the best at multitasking, they are actually the worst. People can't help but multitask and are distracted by devices and screens. They can't help themselves and rationalise to themselves that they are good at it."

As my life went from offline to online, I quickly found myself grappling with a high number of tasks and micro-tasks. To cope, I started multitasking heavily. To begin with, I felt like a dynamo, getting all of the things done. But by the end of the day I had an empty feeling of having done nothing at all. I usually had a to-do list comprising a large number of micro-tasks like responding to emails, publicising an article, and commenting on blogs, and a smaller number of substantial tasks. Every micro-task was written down and ticked off, making my to-do list look like it was very active. But the big tasks just hung around on my list, too big and requiring too much thought to be approached. I found myself searching for ways to avoid them. I found myself procrastinating.

I have always been a procrastinator. As a teenager, before the internet was everywhere, I avoided procrastinating by going to the library to study for my A-Levels. There were still ways to procrastinate, wandering around shops, stopping into places where friends worked for a chat, flicking through magazines, sorting out the junk in my coat pockets, but these were all physical acts. I had to physically do something to procrastinate.

Now, from my chair, from exactly the same physical location where I work, I have boundless opportunity for procrastination. The internet's micro-tasks make it easy to do just one more thing and one more thing and one more thing ad infinitum. Check email, look at Twitter, respond to a message. There's always one more thing to do before I get started on the real task. An email comes in and once I've read it the response is already forming in my head, and though I try to put it aside it keeps bothering me. As an ex-smoker, it reminds me of when I was unable to start work until I'd had just one more cigarette.

Back in 2003, as an avoidance technique for writing essays for my undergraduate degree, I started blogging. I loved blogging. It introduced me to people who I may not have met otherwise and brought me into a like-minded online community. It was also my first experience of the opportunities for hyper-procrastination afforded by the internet. Blogging was social. It was a way of sharing ideas and talking about those ideas. Today blogging is more often used for marketing than for writing and sharing ideas but back then there was just a fraction of the number of blogs and it was associated with geeks and nerds.

It wasn't blogging per se that was the cause of my procrastination. Bloggers don't just write–they are engaged in a conversation and a network. I had a stat counter which I would check obsessively to see the traffic on a new post. For a brief period there was a service called Reinvigorate, a live monitor of your blog that beeped when a visitor landed on your site, giving real-time information about them. That was brilliant. Each beep was a tiny thrill. I not only knew that someone was on my blog but I could see where they were and what browser they were using.

I used Bloglines, a blog-aggregator service, which told me when my favourite blogs had new content. My essay writing would be interrupted as I refreshed the page in search of new content. And there were comment threads. Now frequently cesspools of the internet where people go to troll, comment threads were then places where discussions happened and friendships formed. I checked not only my blog for comments, but those of my friends and the people I followed. I could interact with authors and other readers. These few things made it possible for me to procrastinate; that and playing PopCap games for hours on end.

I remember when Apple launched the iPhone.

"No way," I said to D, "what a pile of crap. Who would want their mail with them all the time?" But it turns out that I did. And so did everyone else.

The internet is no longer contained on my computer on my desk. It is in my pocket. It is increasingly in other objects, those not normally associated with communication–watches, thermostats, lighting, fire alarms. It is part of everyday things. It is an infrastructure. Now, if I want to do anything that involves major levels of focus I have to take active steps to intervene. There is an application called Freedom which blocks the internet on your device, giving you the freedom you need to get work done. When I went away to write for a month I left my smartphone behind, instead having a dumbphone, an old-style mobile phone so that I could just be contacted by text message or phone.

My smartphone creates a low level of mental pressure. As a tool, it can be incredibly useful, but it can also be destructive. With the internet at hand, it's always possible to instantly take action on any stray thought that enters my head. This keeps thoughts on the surface. It allows inconsequential things to break my focus. I can flit from stray thought to stray thought, constantly occupied, taking action on things that may take seconds but that in total mean putting off what I actually want to be doing. It means that there is no space for allowing unformed thoughts to come to the surface. There is nowhere for ideas to grow and breathe. And we become so accustomed to always being switched on that when we do find ourselves suddenly disconnected we are bored.

I sit on a set of concrete steps on the beach near my home. I am writing. This requires focus. But as I do so thoughts flicker in my head that make my hand twitch, anticipating the reach for the smartphone in my pocket. It would be so easy to pull it out and give in, to find answers to whimsical questions thrown up by my mind: what time is low tide today? What stretch of the beach are dogs forbidden from walking on during the summer? Did that guy respond to my email about renting office space? Should I quickly respond to that question about a workshop? What did that article I read last year have to say about procrastination? Will it rain today? Did anyone Like the photo of the beach I posted to Facebook? What should I have for lunch?

Each of these distractions, or interruptions, is a step down the path of procrastination. This is compounded by the fact that my iPhone is a medium for shit that everyone else in the world has to say. I can spend hours just reading stuff that I'm not that interested in.

The things that I do to procrastinate are easy. And the things that I should be doing are hard. It's those hard tasks that dropped to the bottom of my to-do list. Lots of small tasks get ticked off, but the big ones hang around for ages. Task-switching leads to distraction. The longer the period of distraction the harder it is to get started.

A good workday looks like this: get up, write, eat breakfast, sit down, and get straight to work. Work until lunchtime, check email, respond to email, chat a bit. Do a few more hours' solid work. Finish. Slouch.

A bad day looks like this: wake up. Grab phone–check email, Facebook, Twitter. Lie in bed scanning internet for up to an hour. Roll out of bed. Eat breakfast. Check email and Facebook again. Go to my desk and turn on my computer. Open Slack and check for messages. Say hello to colleagues. Half-make a to-do list. Check email. Respond to a few emails. Chat to Australians. Respond to discussions on internal blogs. Check email. Browse Facebook. Get a snack. Do a few housekeeping and admin tasks. Check email. Read the news. Ignore increasing pressure to get started. Respond to a few emails. Start a piece of work. Check Twitter. Click an article. Scan it. Save it to read fully later. Go back to work. Check email. Prioritise to-do list. Make lunch. Eat lunch while chatting to friends online. Decide that since I'm not getting anything done at home I should go out to work. Spend fifteen minutes looking for new headphones on Amazon. Walk to coffee shop. Check email while waiting to order. Check Slack. Like a few things on Facebook. Order a snack. Sit down and surf internet as I wait for food; I don't want the arrival of my order to break my concentration. Look for my notebook with my list but have left it at home. Check email. Answer emails. Browse latest movie news. Say hello to friends based on the east coast US. Order arrives. Get my iPod out and try to find something to listen to. Nothing there. Get out iPhone. Check email. Look for something on Spotify. Feeling uninspired so look at what my friends on Spotify are listening to. Spend a few minutes scanning that. Try a few things, then default to Spotify's Music for Concentration playlist. Check email one more time. Start work. Open text editor. Stare blankly at the blank page. Think about the right phrasing for a Facebook post about how hard it is to start working and the irony of posting about it on Facebook. Post it. Notice someone has shared a BuzzFeed article about people with curly hair. Click. Read it. Ha ha—their life is just like mine! Look back at Facebook. One Like on my post. I thought more people would have liked it than that. Get back to work. Write half a paragraph. Give up on what I'm doing to start something else. Check my email to see what I've got to do. Respond to some emails. Delete a bunch of spam. It's 15:35—the coffee shop is closing in twenty-five mins. No point starting anything. Check Facebook.

Two more Likes. Maybe I should have phrased it a little differently. Pack up my things. Wander home.

Say hi to D. "I've been unproductive today," I moan. "Going to have to work this evening."

Slump onto the sofa. Pull out laptop. Turn on *Pointless*. Say hi to my friends in PST. Chat to them as I watch *Pointless*. Make dinner. Eat dinner. Check email. Go upstairs to get my list. Decide on something that's doable with the television on in the background. Open text editor. Catch a glance of the lampshade that I hate. Search Google for new lampshades. Remember that I need to buy a new travel plug. Go to Amazon. Buy plug. Go back to looking for lampshades. It's always impossible to find one; why does no one make decent lampshades? Check email. Look on Facebook. Only seven Likes. Frown. Look at the cinema listings to find out what's on.

It's 9pm. Too late to start anything really. I keep looking for lampshades, taking a diversion into desks, rugs, and coffee makers. Disappear down a Wikipedia rabbit hole that goes Yellow Magic Orchestra, Haroumi Hosono, videogame music, Commodore Amiga, Dick van Dyke, Mayflower, indentured servants, Wars of the Three Kingdoms, Puritan, hedonism, transhumanism, the world's most dangerous ideas, Martha Nussbaum, Wolfenden report, Michael Pitt-Rivers, Larmer Tree Gardens, Thomas Hardy.

Go to bed. Check Facebook. Check email. Moan to D once again that I've got nothing done. Try to read. Can't read. Play Candy Crush for an hour. Check email. Can't sleep. Toss and turn. Eventually sleep.

That is the worst of days. My days usually lie somewhere in-between, a mixture of getting things done and procrastination. A day of procrastinating leaves me feeling miserable. If I work and get things done I feel engaged. It gives depth and shape to my time. When I procrastinate I just skim along the surface, from bit to bit to bit. When I get work done it's easier to switch off when I'm not working. My mind doesn't buzz with things that need to be done. When I don't get work done it takes over the rest of the day as a possibility, as something I could or should be doing. It means that I can't really commit fully to anything else.

It wasn't long after I started working from home that procrastination really became a problem. One of the benefits to working from home is supposed to be fewer distractions, but when you work on the internet you have infinite distractions. Imagine a giant open-plan office in which everyone is welcome; people stop by your desk, say hi, show you funny pictures of their kids and videos of their cats, say stupid shit you end up arguing with. Not only are people always around, but your open-plan office happens to be in the biggest library in the world – there is always something for you to read, whatever your whim. It is very easy to not do work but to make yourself feel like you are busy. And it's hard to avoid distractions when you work from the rabbit hole. I realised, pretty quickly, that I needed to figure out a way to be productive.

Stuck in my head, I feel that I am the only one who procrastinates to this level but, through the internet, I find solidarity. On the internet, there are innumerable articles about procrastination. There are so many people doing it and struggling with it that they feel the need to write about it on their blogs, websites, and in national newspapers, the best of which is an article by Tim Urban on the blog *Wait But Why*. Urban uses cartoons to characterise procrastination through the struggle between the Rational Decision Maker and the Instant Gratification Monkey. The Rational Decision Maker is the normal part of the procrastinator's brain that is just like everyone else, focused on goals and getting things done. But the procrastinator is plagued by the Instant Gratification Monkey who lives in the present and whom the Rational Decision Maker is unable to keep under control. The monkey wants to spend all of his time in the Dark Playground, "where leisure activities happen at times when leisure activities are not supposed to be happening." This is not a good place to be because the "fun you have in the Dark Playground isn't actually fun because it's completely unearned and the air is filled with guilt, anxiety, self-hatred, and dread." The only thing that can scare off the Instant Gratification Monkey is the Panic Monster, which appears as a deadline is about to approach, giving the Rational Decision Maker a short period of time to scramble through their work and submit it. My paraphrasing does not do it justice.

Productivity techniques, tools, and tropes are not a modern invention. Like so much about our modern attitude towards work, productivity has its roots in Puritanism. Time must not be wasted. We must extract the most value from every minute and every second that we are alive. Max Weber, in *The Protestant Ethic and The Spirit of Capitalism*, writes:

Waste of time is thus the first and in principle the

deadliest of sins. The span of human life is infinitely short and precious to make sure of one's own election. Loss of time through sociability, idle talk, luxury, even more sleep than is necessary for health, six to at most eight hours, is worthy of absolute moral condemnation.

Despite having lost some of its theological overtones, this ethic of productivity is still alive and well. When we are productive we overcome many of our human failings, our tendency towards trivia and laziness. We must extract the best out of ourselves, our time, and our life.

The modern quest for ever more efficiency and productivity can be traced to the late nineteenth century, to the Efficiency Movement, and to the ideas of Frederick Winslow Taylor and scientific management. In the late nineteenth and early twentieth century, Taylor sought to get rid of inefficiencies in production by optimising work methods. He pioneered the time study. This breaks down a job into component parts, timing each part, and then arranging all of those parts into the most efficient working method. Each worker is responsible for a single part of the production process, which is ordered and managed to be of maximum efficiency. Thought about the production process moved from the individual worker to the manager, who is responsible for taking a holistic view and managing all of the component parts.

As part of his studies, Taylor noted that without any extrinsic motivation, workers work at the slowest rate that goes unpunished. They are not working at their optimum productivity. This is no surprise since the time study made the worker just one cog in a machine, removing any sense of connection to the goods that worker produced. He therefore recommended piece rates so that workers were compensated for what they produced, rather than just the hours spent working. This provided an element of extrinsic motivation and encouraged them to be more productive.

Taylor's time studies were often paired with the motion studies of Frank and Lillian Gilbreth, who used recording equipment to film workers' movements and body posture while recording the time. This creates a visual record of work, allowing managers to discern the best way for a worker to carry out a task.

Taylorism was largely unsuccessful in its implementation: managers who tried to implement Taylor's ideas did so without a deep knowledge of them, often focusing on ideas to do with deskilling and monitoring, rather than fair pay and conditions. It was superseded by other management practices. But Taylor paved the way for a set of questions: how can we improve work processes? How can we eradicate inefficiencies? How can we optimise modes of production? It also marked the start of our obsession with recording. To enable analysis and improvement, a time and motion study records both time spent and worker movement. This early use of data for self-improvement contains the germination of today's preoccupation with using data to analyse our time and make sure it is well spent. The best way to improve efficiency and eliminate wastage is to record in minute detail everything that I do and use that data to plan what I am going to do and how I will do it.

Today's manifestation of Taylorism doesn't come in the form of top-down management but in self-help and self-management techniques that help individuals to manage their time better, also known as "life hacking." According to Wikipedia, "life hacking refers to any trick, shortcut, or novel method that increases productivity and efficiency, in all walks of life." It has had particular impact in the tech industry, where people are always looking for ways to do tasks faster and be more productive. There are blogs and communities dedicated to life hacking, like the eponymous Lifehacker.com.

Productivity tools and techniques offer us a sense of control in a world that often feels both fast-moving and without edges. Whereas Taylor and other theorists of scientific management sought to give control to managers by breaking down tasks and making them measurable, our concern today is with regaining control over our own lives. While my day-to-day job might not be predictable, I can make the process of doing it predictable through the use of to-do lists and by blocking out my time. They offer me a way of dealing with information and data when I exist in a highly networked, highly fluctuating world that I have little control over.

Productivity applications have proliferated at a time when the ability of companies to measure performance against traditional standards–time and location–is being diminished. When employees work remotely, it's difficult for companies to know when they are working or what they are doing while they are working. This means that workers become responsible for their own output and productivity.

There are many blogs about life hacking and productivity, a further indicator of just how widespread the malady of procrastination is. The problem with such blogs is that they themselves can become a productivity nightmare. Reading articles about productivity and following their advice makes you feel like you're doing something, that you are taking the right steps towards being productive. You feel busy. You are busy. But you are usually stuck in a cycle of throat-clearing and preliminaries.

I started my own foray into life hacking with one of the most prominent and popular techniques – Dave Allen's *Getting Things Done* (GTD). GTD is a time-management technique in which practitioners record everything they do, clearing away their mind, packaging it up and organising it. This moves tasks out of their mind and ensures that nothing get lost. By shifting things out of your mind and into your GTD system, you can focus on individual tasks, without any anxiety that you'll forget something.

The system behind GTD is outlined in Dave Allen's 2001 book of the same name. I bought the book, read it, and was sold on the idea of externalising and sorting everything that I have to do. What I didn't realise was that it could become an obsession. I am the type of person who enjoys the organisational process in and of itself. I particularly like big taxonomical systems which allow me to sort and categorise things. GTD provides exactly that. It also provides a workflow and process to be followed. For the system to work, absolutely everything that you do must be captured and inventoried, everything from errands, to emails, to feeding the cat and doing the shopping.

The system has five main steps. First, empty every thought, idea, and task that you have in your head into a "collection bucket." This could be a notebook or a document or a piece of software. This clears everything from your head, capturing it externally. Second, clarify everything you've written down so that it becomes a task that you can take action on; so, for example, "write blog post" is broken into actionable steps such as do research, write first draft, edit, publish, publicise. Third, organise and prioritise everything–assign due dates, categorise it, set reminders. Fourth, reflect on your list and see what your next action will be. Since you've just prioritised all your work this should be easy. Finally, engage and get on with your work.

Allen recommends using a notepad and paper, but even before I got started I knew I wanted an app. There is a plethora of apps for GTD. In the end, it doesn't really matter which one you use since it's all about the categorisation system and list, and most to-do applications have tagging and sorting features that can be used with GTD. I tried Wunderlist, Any.do, Todoist, Things, Omnifocus. I loved the first few steps, the process of writing everything down, clarifying it, and prioritising it. Once I had done that in an application I would "engage," carrying out my work for a while, but some niggling problem with the application got in the way and I'd start again. My enthusiasm eventually petered out. I stopped categorising everything. My approach became scattered. I would ignore everything on my to-do list, not really arsed about it anymore, and after weeks spent learning and implementing GTD I was back where I started.

I don't have the staying power for a long-term organisational method like GTD. You need to be obsessional and bordering on OCD, let the system shape your life. And maybe that's part of the problem. I had to learn a whole new thing just to do the things I needed to be doing. It was getting in the way of real work. The act of learning it made me feel productive while all I was doing was messing around.

I went back to the dangerous predicament of swinging between productivity and non-productivity, a precarious state for the procrastinator. I installed an app called RescueTime on my laptop and my phone. RescueTime tracks everything that you do, so you can see how much time you spend actually working and how much time you waste, with the aim of helping you to cut down on time-wasting. I hoped that once I saw how much time I spent on Facebook I'd be so horrified that I'd work harder. I switched it on and left it running in the background.

My next productivity technique was the Pomodoro Technique. To date, it has been the most effective. Even today I go through phases of using it and not using it. When I'm having slumps in productivity I think back to those anxiety-free days of productivity bliss. The Pomodoro Technique was invented in Italy the 1980s by Francisco Cirillo, who used his tomato-shaped, or pomodoro, kitchen timer. In my house I use a chicken-shaped timer, so when I'm working while using the technique it's called "chicken time."

What appeals to me about the Pomodoro Technique is that it is an approach to time-management that tackles the problem of time. The premise is that we experience the passing of time, or *becoming*, and this causes us anxiety. Two hours have passed and I haven't completed what I needed to do. It's already Wednesday and I haven't started on my big task. By trying to measure ourselves against the passing of time we become anxious, diminishing our ability to act. The technique seeks to change how we view time. Instead of approaching time as an ongoing process, it breaks work down into chunks of twenty-five minutes. This enables you to experience time as a sequence of discreet events, thus making it more manageable.

To practice the technique you decide on a task, set your timer for twenty-five minutes (a unit of time called a pomodoro), and work until the timer goes off. If anything comes up or pops into your head you write it down, externalising it and saving it for later. If you are completely interrupted you end the pomodoro.

By doing this, you no longer meander through the long, undifferentiated stretch of time that is your day. You focus on tasks that are compartmentalised into chunks of time. This helps to address the new tension that has emerged in our lives: that we are brought up to be time-oriented, but working in a task-oriented manner. It's this tension that perhaps explains why I found it so difficult to sit down, shut up, and get my work done. Like most other people, I was used to my time being structured and defined by something external to me. I was subject to the time-discipline of the bell at school, the timetable, the rota, the nine-to-five. Now I was without that and I was lost. I needed something to introduce some discipline into my life.

E.P. Thompson, in his article "Time, Work-Discipline, and Industrial Capitalism", discusses the difference between task-orientation and time-orientation. Task-orientation is associated with work pre-clock, in peasant and rural societies, and any culture that disregards clock time, judging time based on factors such as the rise and set of the sun. Despite being found in tribal and rural communities, task-orientation has many similarities to the task-driven nature of online work and freelancing. Thompson outlines three aspects of task-orientation, all of which resonate with my own work: that work is approached through necessity, not because of the time on the clock; that people who work in a task-oriented manner have a more blurred boundary between "work" and "life"; and, finally, that people who work according to a clock find that those who are task-oriented lack in urgency.

Pre-Industrial Revolution, when work was done on a domestic or small-workshop scale, there wasn't much need for task synchronisation, so people worked in a task-oriented manner. The amount of work a labourer did depended on what was needed: to produce a certain number of pairs of shoes, or weave a length of cloth. Many craftspeople and labourers did multiple different tasks. The diary of a farming weaver from 1782–83 notes many of the tasks he undertook. These include weaving, threshing, harvesting, mending clothes, laying up a coal heap, sweeping the roof and the walls of the kitchen, laying muck midden, churning, ditching, gardening, preparing a calf stall, cutting down trees, helping birth a calf, going to buy medicine, working with a lathe, writing letters, picking cherries, working on a mill dam, "jobbing" with a horse and cart, attending a Baptist association, and going to a hanging. The work pattern was a mixture of intense work, idleness, and focus on many different tasks. The pattern is not all that different to many freelancers. It's unlikely that a freelancer would be laying a muck midden, birthing a calf, or churning, but the list of tasks that they carry out in a week is equally long and it is often marked by bursts of intensity alongside idleness.

The Industrial Revolution and the gradual move to industrial capitalism saw the introduction of time-discipline on labourers. The Law Book of the Crowley Ironworks is an early eighteenth-century 100,000-word civil and penal code which Sir Ambrose Crowley used to discipline his workers. It outlines the hours for which a worker would be paid, minus deductions for "being at taverns, alehouses, coffee houses, breakfast, dinner, playing, sleeping, smoaking, singing, reading of news history, quarrelling, contention, disputes or anything forreign to my business, any way loytering." Time is kept by a Monitor and a Warden who keep a timesheet for each employee, and who monitor the time that workers spend on tasks. This type of monitoring and timekeeping is a precursor to the widespread adoption of time-discipline in the Industrial Revolution.

The introduction of time-discipline in factories was accompanied by pamphlets which sought to impose "time-thrift" on domestic and social life. A pamphlet written by Rev. J. Clayton and distributed in 1755 entitled *Friendly Advice to the Poor; written and published at the request of the late and present Officers of the Town of Manchester*, gives the following friendly advice: "If the sluggard hides his hands in his bosom, rather than applies them to work; if he spends his Time in Sauntring, impairs in Constitution by Laziness, and dulls his Spirit by Indolence," then all he can expect is poverty. This is one of the first moralistic outcries against older customs, holidays, and ways of being. As well as imposing time-thrift on the poor, another group was targeted:

children. Children were expected to go to school, observe hours, be punctual. If children were poor they were, from the age of four, to be sent to workhouses where they would get two hours of schooling a day and for the rest of the day be engaged in manufacturing. The school was, and is, the place of disciplined time. As soon as the schoolchild reaches the school gates they are subject to the time-discipline of the school. They answer to bells and timetables that tell them where to be and when.

As the Industrial Revolution progressed, time-discipline and time-thrift were imposed on every aspect of life: school, the factory, the workshop, the poorhouse, domestic life. In some factories, workers were expected to work to a clock but they were not permitted to see that clock so they never knew how much time they had spent at work or how much they had left.

The shift from task-orientation to time-orientation was not just due to the imposition of time-discipline by factory owners and moralists. Alongside this was the spread of Puritanism and the Protestant work ethic. With the Protestant work ethic, time is a precious commodity and not something to be wasted. An article by Oliver Heyword in *Youth's Monitor* (1689) reads that time "is too precious a commodity to be undervalued... This is the golden chain on which hangs a massy eternity; the loss of time is insufferable, because irrecoverable." John Wesley, one of the founders of Methodism, wrote of time:

See that ye walk circumspectly, says the Apostle... redeeming the time; saving all the time you can for the best purposes; buying up every fleeting moment out of the hands of sin and Satan, out of the hands of sloth, ease, pleasure, worldly business...

One of the most significant tools for the imposition of time-discipline was the timetable. A timetable is a way of carving up time into strict blocks, telling you what to do and where to be at specific times. It reinforces habits through repetition – every Tuesday at 12:30 PM I go to the gym – this happens with such regularity that I don't even need to think about it anymore. The strict adherence to a timetable has a long history, beyond the Industrial Revolution. Its roots can be found in monastic traditions. Timetables provided a sense of order and of moral rightness, interweaving time for prayer and worship throughout the day with the monks' other more mundane duties. They were later taken up by other institutions – schools, poorhouses, workhouses, prisons, and hospitals – and became essential in the workplace. A list of rules distributed to the employees of the Foundry and Engineering Works of the Royal Overseas Trading Company in Berlin in 1844, in use more than a century after the Law Book of the Crowley Ironworks, shows just how strict the adherence to time had become:

(1) The normal working day begins at all seasons at 6 PM precisely and ends, after the usual break of half an hour for breakfast, an hour for dinner and half an hour for tea, at 7 PM, and it shall be strictly observed. Five minutes before the beginning of the stated hours of work until their actual commencement, a bell shall ring and indicate that every worker employed in the

concern has to proceed to his place of work, in order to start as soon as the bell stops. The doorkeeper shall lock the door punctually at 6 AM, 8:30 AM, 1 PM and 4:30 PM. Workers arriving 2 minutes late shall lose half an hour's wages; whoever is more than 2 minutes late may not start work until after the next break, or at least shall lose his wages until then. Any disputes about the correct time shall be settled by the clock mounted above the gatekeeper s [sic] lodge. These rules are valid both for time- and for piece-workers, and in cases of breaches of these rules, workmen shall be fined in proportion to their earnings. The deductions from the wage shall be entered in the wage-book of the gatekeeper whose duty they are; they shall be unconditionally accepted as it will not be possible to enter into any discussions about them.

Workers had their wages docked even if they got ready to leave before the ringing of the bell. The foreman or the factory owners may be the explicit symbols of power, but it's time that's really at stake. It is time that shapes the workers, trains them, disciplines them. In the timetable, clock time is used as a method of control. It is internalised through repetition, rhythm, and habit. It's often said that it takes thirty days to create a habit. Recent research suggests this to be more like 66 days, but what matters is this: with the correct training we can assimilate anything to become part of the natural fabric of our lives.

Through the Industrial Revolution time-discipline was imposed on the masses. It came from all sides: the introduction of clocks, watches and bells, time sheets and timetables, moralising about the importance of time, the suppression of old ways of working, through schools and in factories. Thompson remarks that by the 1830s and the 1840s a commonly noted difference between English and Irish workers was that English workers, unlike their Irish counterparts, were regular, methodical, and unable to relax. In today's society we are so used to time-discipline that there is nothing remarkable about bells and timetables, we expect to work a nine-to-five, and for many people there is a clear demarcation between the time of work and the time of life.

The internet has broken down this clear distinction. Even people who work a nine-to-five job often find themselves doing additional work at evenings and weekends, responding to emails or finishing up pieces of work that they couldn't get done in the office. Freelancers, contractors, and the self-employed find that nine-to-five no longer applies. Our work is task-oriented, having more in common with pre-industrial and rural societies than the timetables of the Industrial Revolution. Work is a list of things that need to be done, not a clearly demarcated period of time. And yet as a post-industrial society we are highly encultured to be time-oriented. We spend fourteen years at school, answering to bells and timetables. When we leave school most jobs are nine-to-five or shift work, defined by the amount of time that we spend on them.

When I quit my job I went from being time-oriented to task-oriented. I no longer found my time subject to external constraints. At first this felt like freedom, but the tension between being brought up time-oriented and suddenly being thrown into task-orientation undoubtedly contributed to both my multitasking and my procrastination. I was unable to properly cope with undifferentiated stretches of time. Like most freelancers, I quickly discovered that I needed to impose time-discipline on myself. And to do this the most obvious tools were familiar. Timetables, lists, and techniques to ensure that time is best spent. We must extract the most from our time, become more efficient, better workers, make the best use of our resources. With task lists and timetables we allocate time, but more and more we use technology. An application like RescueTime is the modern equivalent to the time-and-motion study. It tracks everything that you do: email, meetings, web-browsing, social media, writing, coding, or whatever it is that you do on your computer or phone. The website says "RescueTime helps you understand your daily habits so you can focus and be more productive." With it, I can scrutinise my time and identify areas where I can improve efficiency.

This drive to be ever more productive is synonymous with what the philosopher Michel Foucault calls "disciplinary societies," in which, through ever more detailed arrangements of time, individuals "tend towards an ideal point at which one maintained maximum speed and maximum efficiency." Unlike the Berlin factory, where an arbitrary timetable was imposed from the outside, an application like RescueTime facilitates a system of control that is self-generated from what you do, and it is a way to make you, the individual, more productive, happier, better. We train ourselves to be better operators, to have maximum efficiency. We arrange time to make the best use of it, each minute accounted for and tallied up. Work time must be productive and, most importantly, profitable. Time not well spent is time wasted.

The Pomodoro Technique, with its focus on time, helped to alleviate this tension between being raised to be time-oriented but working in a task-oriented way. It structures undifferentiated stretches of time so tasks can be completed. My success at the Pomodoro Technique lasted about six months. It was a period of hyper-productivity, but, even though I was doing fine, I started looking around for new apps to help me do it better. Despite the temporal barrier created by each pomodoro, I still found ways to procrastinate. I put off starting so I could only get through ten pomodoros, then six, then I'd start one and give up halfway through and that was it.

My adventures in productivity culminated with timeboxing. Timeboxing is similar to the Pomodoro Technique in that it approaches time by boxing it into sections. You divide your time into boxes and allocate tasks into those boxes. It's similar to a timetable, in that it breaks up your time into pre-allocated chunks. This, no doubt, was the moment when my obsession with productivity slid into madness.

I sat at my computer and started a blank Google Spreadsheet. Here was where I would carve up my time, here was where I would allocate it properly and ensure that none of my time would be wasted. Everything must be accounted for, every element of my life divided across the cells of a spreadsheet. I would create the perfect timetable for my life.

I spent a full day of deep focus creating a spreadsheet to timebox effectively. I worked hard, allowing no distractions or interruptions. I created a cover sheet that listed my to-dos, and a taxonomical system for organising them. This I broke down into:

- Type: Work, Home, Personal Goal, Health
- Time: Micro, Short, Medium, Long
- Priority: Low, Medium, High
- Status: Allocated, Waiting, On Hold, Complete
- Project
- Activity
- Notes

For each weekday I created a tab – Monday, Tuesday, Wednesday, Thursday, Friday. On each sheet I broke the day down into time slots which I allocated to different tasks, ensuring that my day had a reasonable spread of tasks associated with Home, Work, Health, and Personal Goals. These are all important after all; I must work but I must make time for home, I need to take better care of my health, and to be a well-rounded person I must have personal growth. I made the spreadsheet calculate exactly how much time I spent on each area of activity. All time must be allocated. None should be wasted. I must spend exactly the right amount of time every day on each aspect of my life.

I used it for one week only. Monday was broken down meticulously into breakfast, email, writing, gym, pilates, lunch, micro-tasks, house admin, event-scheduling, picking up parcels, dinner, free time (one hour), writing, stretching, bed. By Friday all I filled in was writing, breakfast, gym, lunch. Just thinking about it now gives me fatigue. It sits in my Google Drive, a manic and crazed object, the perfect storm of procrastination and productivity. Its madness is in its perfect organisation of my entire life. It puts boxes around all of my time, an accountant. It represents, for me, the thing that I thought I had escaped. I had become a Puritan, I had applied Taylorism to my entire life, I was no better than a factory line manager. I had become my own boss.

Worse, I was a petty, bureaucratic boss. One of those bosses who wants every second to be accounted for, who looks over your shoulder to see what you're doing, who wants you to check in and check out. And I didn't just apply it to my work, I had applied it to everything. I had internalised all of the moralising about time well spent, assimilated what I hated, and I had created a tool for making myself submit.

I had no one to tell me what to do so I disciplined myself. I needed an object of oppression, something external that structured my life. I needed my own Crowley Law Book, my own list of rules. Lying in bed in the morning there's no one to tell me off for being late to work, but a voice in my head repeats "you'd better get up now." Even if I don't have much to do it is there, nagging me: "time for work, time for work." The voice in my head only lets me take thirty minutes for lunch, makes me feel guilty for taking a walk, tells me that I must work late if I don't get enough done during the day. The Puritan inside me is horrified by my procrastination, is obsessed with time-keeping and time-tracking: this time for writing, this for exercise, this for spending time with family, this for friends, this for DIY, this for socialising, this for stretching, this for reading, this for leisure, this for watching TV, this for cleaning, this for showering, this for shopping, this for eating, sleeping, fucking.

It's a running joke amongst freelancers to say

"my boss is an asshole," while we work long hours, battle with procrastination, take on more and more work. Thoroughly time-disciplined, we fight with the ephemeral nature of the internet, engaging in ever more cruel self-berating. That spreadsheet, the zenith of my productivity efforts, is its most perfect representation. After I created it, I realised what it meant and what I had become: I ended my obsession with life hacking and productivity. I work with my notebook and my list. I have stopped charting and accounting everything that I do. I have turned off RescueTime. I get done what I get done and I have, for now, escaped the worst boss I ever had.

CHAPTER THREE

NEAR YET FAR

I had been working online for less than a year when the arguments started. D and I were still living in a small house in St Margaret's at Cliff, near Dover. In the daytime I worked in my office and in the evenings I sat on our cramped red sofa, laptop on my knee, working until I went to bed. I was still in the haze of figuring out how to get maximum productivity out of every hour of my day: this had an impact not just on me, but on D and our relationship. Every hour of my day was monetisable, but every hour of his was not. For me, idle time was time wasted; for him it was downtime from a job he didn't enjoy. Paid by the hour I wanted to work every moment possible; paid a salary he wanted to work as little as he could.

It was D who encouraged me to take that first step and quit my job, and D who provided the support both while I searched for those early jobs and while I made changes throughout my online career. We both believe that being in a relationship should be liberating, not constraining, and that our roles are to support one another to become more than ourselves, to experience life fully, rather than hold each other back. But that doesn't mean that he was immune to my attitude towards work, and the greater my tunnel vision, the more the tension increased. It began with him asking if I was planning to work all evening again, and me pointing out that it didn't make much difference for me to be working since all we did was sit around watching TV anyway. We would sit together, me with my laptop, he with a book, the television burbling on in the background. If he tried to engage me in conversation I responded with a half-answer, not really paying attention to what he was saying and resentful that he was breaking my train of thought.

I felt a distance between us: I was there with him but I was never really present. I was always looking at a screen, whether that screen was my laptop or my smartphone. There was a permanent intrusion on our life and our relationship. My work had become part of our home. Before I started freelancing, D and I left the house, went to our jobs, and returned home to be with one another. Home was a protected space, somewhere that we could go to recuperate together. It's true that sometimes we would "take work home" in the form of stresses and anxieties, but we were able to keep work itself at bay and out of the household.

When I began to work online, the division between these two distinct areas of our life broke down. I can't deny the many advantages to working from home: there's no commute, you can focus on getting your work done, cook and eat meals at home, and avoid office politics. But by bringing work into the home I had destroyed the space that was just for the two of us, our space and time to be together. D still inhabited that space, expected it when he got home from his day-to-day job, but I was no longer there, I had abandoned him, turned away towards a virtual world which had no clear demarcations in space or time.

Boundaries, whether they be temporal, physical, cognitive, relational, or emotional are an important way to manage the different aspects of a person's life. The traditional boundary between work and home allows a person to move between one zone and another, keeping conflicts and stresses clearly demarcated as belonging to one area, giving individuals distinct spaces in which to recuperate from stresses occurring in another zone. A person recovers from work at home, but equally they might recover from home at work. Remote work eradicates that easy physical distinction. Individuals must create new boundaries, ones that may be temporal (marking out times of the day to work), spatial (creating zones for work, whether that's at home or in a co-working space), or cognitive (creating habits or practices that allow the person to switch modes). While remote work might allow a person to easily make the physical switch between work and family, it also increases the possibility of interruption across roles (children or partners interrupting work/emails and calls outside your normal working hours). Research has found that people who telecommute regularly are more likely to work long hours; telecommuting, far from being a straightforward employee perk, is a way for companies to account for the fact that the amount of work expected of someone is far greater than they can do in a standard forty-hour week. It also raises the expectation that employees will be able to work in the evenings and at weekends.

This is a particular problem for telecommuters and remote workers, people who work from home all of the time. But it's become more and more of a feature of every job. Most people have access to their emails at home and may send out electronic correspondence in the evenings or on weekends. It can be hard in any job to completely separate work from the rest of your life. While there are many who laud this blending of the distinction between work and life, these are often people whose jobs are synonymous with their lifestyle. For everyone else, it places unnecessary strain and creates the expectation of an always-on culture. This is not to the benefit of employees. A long-term study of employees from 1989 to 2008 found that people who do additional work from home outside of their normal working hours are not paid overtime for this work, thus they are carrying out free labour for their employers. In contrast, people who stay in an office and do overtime are compensated by pay and will see an annual increase in their wages.

Telecommuting became part of common vernacular in the 1970s, when Jack Nilles, an engineer working at NASA, set up a telework project with thirty employees. The experiment was intended to address problems that arose due to the oil crisis. Nilles' project looked at ways to bring work to workers, rather than the other way around. This would alleviate traffic problems and reduce energy consumption. Shortly afterwards, the US government set up federal programmes to investigate the feasibility of telecommuting, and by

1997 ten thousand federal government employees were working remotely.

Technological advances have made it increasingly easy for individuals to work from home. Personal computers were introduced into homes in the 1980s, the 1990s saw an increase in the use of laptops and mobile phones, and the twenty-first century has seen the rise of smartphones and other networked devices. These technologies have increased in power and decreased in price, and the revolution of the internet has brought down any communication barriers. Today, I have access to the same data and tools at home as I would in any office. As well as technological changes, telecommuting has been made easier by the shift in industries from manufacturing to information, making it less necessary for workers to be in the same place.

Telecommuting is becoming more and more prevalent. A 2011 poll from Ipsos/Reuters that surveyed twenty-four countries found that 10% of employees work at home every day, and one in five telecommute frequently, with the practice particularly common in Mexico, Indonesia, and India. Global workplace analytics reports that between 2004 and 2014 the number of employees telecommuting has increased from 1,819,355 to 3,677,061 ¼ a growth of 102%. These statistics just include those who work for a company. As of 2014, around 22% of self-employed people work from home.

Telecommuters are as varied as traditional workers, working in many different styles and industries. There are telecommuters who work for a traditional company but who carry out their duties from home, there are people who run their own home businesses, there are people who just work part of the time at home, and increasingly there are members of remote teams–these are usually fully distributed with team members interdependent on one another for work product.

This increase in telework has inevitably affected home-life dynamics. When one person spends their day at home and the other spends their day out at work, each person has a different experience of the home. This change in the experience of space can alter the dynamics of a relationship–one sees home as a space for rest and recuperation, the other a place of business. This is complicated by the fact that the person working from home doesn't always look like they are doing work. When you wear pyjamas all day, perch on a sofa, and line up cups of tea, it doesn't outwardly look like you are doing a lot. It's very different to the person who has to get out of bed, take a shower, maybe put on a suit, and go into the daily grind of an office for 9 AM. It can be difficult for the person who goes out to work to understand what the remote worker does, how, quite quickly, their work becomes so entangled with their life that it can be hard to know where one ends and one begins, that it's not easy to switch off when your home is your office and your office is your home.

This can lead to an imbalance in the amount of housework taken on by each partner, although it is usually female telecommuters who take on more work than their male counterparts. One piece of research found that while husbands who make use of flexible work arrangements did not find their family life interfering with their jobs, wives reported increased strain at work due to family interference after one year of working from home.

There has been increasing interest in how telework and remote work affect family relationships. Researchers talk about work-family conflict, a multidirectional strain that can cause stress both at work and in the home. There have been numerous studies on the interference between family and work and vice versa. One study found that the more an individual telecommutes the less their work interferes with their family but the more their family interferes with their work. Another found that the length of time that a person has teleworked for affects the level of work-family conflict. Greater work-family conflict is experienced by telecommuters who have worked from home for less than a year, and the level decreases in employees who have telecommuted for more than a year, suggesting that over time a remote worker becomes better at managing their time and negotiating that relationship between work and home.

The factors that can lead to work-family conflict for a remote worker are myriad, and they contribute to exhaustion and higher negative stress levels for the individual. Due to the co-location of work and home, there are the constant reminders of home while at work, and vice versa. If a person is experiencing problems and anxieties in either of these domains they have no place that they can escape to. This gets in the way of recovery and recuperation from stresses, depleting any energy that might be required to resolve the conflict. It is difficult for anyone who works from home to distance themselves from conflicts that are occurring in either place. If D and I had an argument at home there was nowhere that I could go to escape from it. Even if he went to work, I was stuck at home with all the reminders and tensions in the environment.

Other problems arise from the fact that remote workers who work unusual hours are often unable to fully engage with the rest of the family's routines, and family members' lack of understanding of an individual's work leads to constant interruptions. More recent research has found that remote work from home can increase strains and stress for individuals who are already experiencing work-family conflict; if people are having problems in their home situations they can be exacerbated by telework and the individual may suffer more because of them. This means that managers need to have an awareness of what is happening in a person's home situation, which is not always practical in fully remote teams.

The way that remote work affects a relationship will differ from couple to couple, depending on the nature of the people, which partner is working from home, and the type of work that they do. Not everyone will experience the problems that I did, although when one person is using the home as a workplace it will inevitably change the dynamic. In some relationships, for example, the partner who goes out to work resents the one who stays at home. This wasn't our problem. He didn't care that I was at home while he was at work. He cared that I spent so much time on my computer rather than spending it with him, even if spending time with him just meant watching a movie or going for a walk. And certainly he was frustrated with the way my work was taking over every aspect of my life.

I realise now that I was disappearing from his

life even though I wasn't going anywhere at all. By always being distracted, I had become the shadow of someone who existed somewhere else. It was hard to ever switch off so I was always distracted. Even when we were together we weren't together. I was looking at my computer or checking my email on my phone or looking at social media. He complained that we never did anything together, I complained that we were together all of the time, he complained that didn't mean I was actually there. But he wasn't the only one getting annoyed: I felt unhappy that he got to leave the house every day, that he had opportunities to make face-to-face relationships, and that he could, if he wanted, go for a coffee or a drink with them. I was excluded from that. Since we had moved to Kent, he had managed to create a real, in-person life for himself. I had not. And so I created a virtual one, one that was as real as his, just dispersed all over the world.

I found the intrusion of the internet difficult to control. I checked my email and messages at all times of the day, responded instantly to push notifications on my phone. The worst moments were always late at night. They still are. I check my email just one last time. I reach for my phone, refresh Gmail, and there is a work email. As soon as I have opened it there is a mental pressure. Small things are just annoying, a buzz that requires my attention but I want to ignore. I refuse to get up and deal with it so I lie there, composing a response in my head, even if it is for the most trivial of things. Worst, though, is the difficult or upsetting message: a negative message from an employer or a peer, something said thoughtlessly, an email that has wound me up. Then I toss and I turn, composing a response in my head, revising and refining, trying to get to sleep, trying to forget it, unable to shut it out. I sleep shallowly, waking regularly to find myself thinking about the problem once again, knowing that the middle of the night isn't the time to respond to anything, feeling trapped and upset in the darkness and waiting for morning so I can deal with it. By the morning I am usually exhausted and find it impossible to do anything rational.

These encroachments happened all the time. Not that D was immune to it–he would equally check his email before bed only to read something that infuriated him and prevented him from sleeping. But with me it was constant. I was never able to escape from work–it was always there, an ever-present demand, an intrusion into my life, our life. I told D when these nighttime intrusions happened, but soon he got infuriated; why didn't I just stop checking my email? I stopped telling him, and when it happened I just seemed grumpy, silent, and pissed off for no apparent reason.

It wasn't just my relationship with D that was affected by the migration of my life to the virtual. We moved to Kent for D's job, and in that time I had gone from a job that I disliked with little human contact, to an online job I loved with lots of virtual contact with other people but nothing in person.

Ever since I first connected to the internet I have made friends online. The first person I chatted with regularly was an American on AOL called IrishMike. We chatted in public chatrooms and on AIM. This was back in the day when chatrooms were punctuated by the question "a/s/l" (age/sex/location). Since then, I've been part of different web communities, via chatrooms and blogs. In the early days, when I was at school, I kept it to myself, not wanting to associate myself with the stereotypically geeky internet user. People generally didn't understand the potential of the internet back then. Why would you want to make friends online when there were plenty of opportunities for making friends in the real world? If you had to go online to make friends it must mean that you couldn't make actual friends. Everyone online was a weirdo. Geeks and emos, sat in their rooms with their RPG posters and their fantasy books, or creepy old blokes who just wanted to get off on talking dirty to you.

People who thought that missed the point. Connecting online meant that you could talk to people from all over the world, people who had more in common than the contingency of physical location. We created little pockets for ourselves, communities of the like-minded where we could express ourselves. We could choose who we wanted to be online, be our best selves. I used to sit all night on my parents' old PC, chatting to people all over the world.

Today, people meet online all of the time. It's unremarkable for couples to meet on Tinder or Match. com. It's not geeky for people to have a partner who they met on the internet. It's become normalised. It is, after all, just a communication tool that enables people to connect.

When I started working online I was already quite comfortable communicating through a screen. It wasn't hard for me to quickly make friends. I was surprised and a little relieved that something I'd been doing since I was a teenager had become so normal.

Invariably, conversations with colleagues moved beyond work. One of the magical things about the internet is that you can use it to communicate and make friends with people all over the world. In those early days of working online I made friends in the US, in Australia, in Eastern Europe, and now my network spans the globe. I can go to most countries and find someone in my extended network who will show me around, help me get grounded.

With burgeoning friendships and a desire to make money, I spent more and more of my time online. My life online contrasted with what was happening offline. Online I talked to people all over the world, to people from different backgrounds, with different experiences and viewpoints, from different cultures. Offline there was just D and I: our everyday lives involved being at home, going for cake in a local tearoom, and the occasional trip to the cinema. We had moved to Kent for his job, and all of my real-life friends were in London and Coventry, Suffolk and Newcastle. I had to travel to see them. There was no one I could just pop out to see; instead I popped online. Being on the internet was meeting my need for both a job and a social life and I didn't even need to leave the house. I had friends available at all times of the day. In the morning I'd chat to people in Asia-Pacific, through the day there were friends from Europe, after lunch people on the east coast US, and late in the afternoon those in Pacific US.

Now, I see how easy it is to get sucked into the online life. It takes so much less effort than picking up the phone, talking to someone, going out. It's easy to slip into patterns that mean that our offline lives are neglected in favour of the virtual world.

"Are you still working?" D would often say.

"No, just chatting." I'd respond, not lifting my face from the screen.

My expanding online friendship group was a way of combatting my growing sense of isolation. I may have been cultivating a diverse group of friendships online, but at home I was lonely. All remote workers are bound together by the simple fact that we experience at least some level of social isolation. This is one of the biggest challenges of remote work. In a poll of remote workers carried out by Ipsos/Reuters, 62% of respondents said they find telecommuting socially isolating. Loneliness and social isolation can lead to higher job turnover and poorer productivity. Some people love the idea of the isolation, but for others it results in them feeling cut off from the rest of the world. Most employees report that face-to-face interaction is the most important form of communication for maintaining workplace friendships. Unless active intervention is taken, working remotely can result in the inability of workers to create lasting and meaningful friendships amongst their colleagues.

Despite my preference for spending long periods of time by myself, I have definitely found loneliness to be one of the most difficult aspects of remote work. D and I moved to Kent around the same time as I started working online. The intensity of those first few years of working online meant I didn't put any effort into making friends where I lived. I spent most of my time at home, working on my laptop. I didn't need to make friends because I had friends all over the world. But there were times that I wished for someone nearby, a circle of friends who I could go out with for a drink at the weekend, or someone who I could go for a coffee with. I wasn't really sure who these people were or where I would find them.

There is undoubtedly a distinction between online and offline friendship. Relationships online are instantaneous and low-risk. When you enter into a conversation with someone in chat you only give as much as you want to give them. The anonymity afforded by the internet can be a comfort for people who find it a place they can open up and interact with others. It can allow introverts to find likeminded people and to form friendships that they would find difficult in person. It can also provide a platform for people who have been traumatised to safely share their experiences and speak to others. Some people find that through the internet they can form more authentic and meaningful relationships than they would be able to offline.

However, the decision about how much one shares gives all online interactions, whether they be public broadcasts on social media or private messages in a direct message (DM), has a performative element. I choose who I am online, holding back the sides of myself which are boring or distasteful or that I think people won't like. I can self-select my best self, or what I see as my best self. I can present myself as I want to be: funny, smart, thoughtful, someone who isn't anxious, who doesn't care what people think of her, someone who doesn't swing between confidence and uncertainty with the regularity of a grandfather clock. I can hide my misanthropy, my disinterestedness, my capacity to be sharp and cutting and mean. I realise that sarcasm doesn't always carry online so I have taught myself to use it rarely, only with people who I know well in an offline context. I know that when I am tired or in a bad mood I find it harder to maintain the performance so it's better, at those times, to step away from the screen. Beyond that, I am much better at articulating myself through writing than through speaking, so I come across as far more eloquent than I really am.

I am not alone in this. The social networks that form the basis of many online relationships are marked by their positivity (Likes, Favourites), and we perform to get the social approval that we all crave. We hide our vulnerabilities and share the parts of our lives which are "awesome" and "amazing." "I'm sick of Facebook," a friend going through a rough patch said to me recently. "All of these people and their happy photographs and their holidays and their new houses and cars and their smiling children and everything I've got is shit." But all of his photographs were the happy, smiling photos too–there was nothing that betrayed what was really going on in his life.

In her book *Reclaiming Conversation: The Power of Talk in a Digital Age*, Sherry Turkle quotes a colleague, Sharon:

> I spend my time online wanting to be seen as witty, intelligent, involved, and having the right ironic distance from everything. Self-reflection should be more about, well, who I am, warts and all, how I really see myself. I worry that I'm giving up the responsibility for who I am to how other people see me. I'm not rigorous about knowing my own mind, my own thoughts. You get lost in your performance. On Twitter, on Facebook, I'm geared toward showing my

best self, showing me to be invulnerable or with as little vulnerability as possible.

This extreme performativity puts us in a permanent state of being-for-others. Who I am only matters as a reflection back of the gaze of other people. We constantly second-guess ourselves in order to appear as we think others wish to see us. Our selves are constituted for others, whether that's to attract their approval, their adoration, or their scorn. When we operate primarily within the performative medium of the internet it's easy to lose sight of who we are for ourselves.

As Turkle points out, this tension between who we really are and the performance of our best selves can lead to depression and social anxiety. It doesn't just happen in social media but in direct messages as well. In any DM conversation, there is a level of mediation beyond face-to-face conversation. This mediation gives me time to articulate thoughts and select the information about myself I want to give away. A client once told me that he always takes emoji use at face value, trusting the person to properly indicate how they are feeling, but this view takes the expression of human experience at face value. Even in the offline world people disassemble about how they are feeling, and online the use of emoji makes this even easier. I can use a smiley face to signal to someone that I am happy when in fact I feel dreadful; one look at my face would tell the other person just how awful I feel. In a space that is marked by its positivity, it's impossible to trust that people are able to tell us when they are feeling bad.

There's a short section of David Foster Wallace's novel *Infinite Jest* which describes, in his fictional world, the move from telephony to videophony and back again. The invention of videophony causes mass excitement and people quickly have it installed in their homes. Soon they discover that without the privacy of audio-only communication, they have to give their full attention to a conversation. People also find themselves unhappy with how they look on the video screen. This kicks off a new mask industry: the first approach to masking is using the screen, a high-definition composite of the best aspects of yourself, but this is quickly supplanted by a more cost-effective solution – a physical mask formed from the person's enhanced facial image. Soon people are demanding ever more refined and aesthetically enhanced versions of themselves:

> Most consumers were now using masks so undeniably better-looking on videophones than their real faces were in person, transmitting to one another such horrendously skewed and enhanced masked images of themselves, that enormous psychosocial stress began to result, large numbers of phone-users suddenly reluctant to leave home and interface personally with people who, they feared, were now habituated to seeing their far-better-looking masked selves on the phone and would on seeing them in person suffer (so went the callers' phobia) the same illusion-shattering aesthetic disappointment that, e.g., certain women who always wear makeup give people the first time they ever see them without makeup.

As the technology develops, videophony provides full-body images and masking techniques become ever more elaborate, taking over a person's entire 2D body image. Eventually, people are using "Transmittable Tableau," high-resolution images of an entire scene which are transmitted instead of the video of the person you are talking to. These are effectively a lens cap, which screen your actual life and transmit your ideal one. Behind the tableau, they are doing exactly what they did on the telephone.

Beyond the hyperbole, the world imagined by Foster Wallace is not dissimilar to our own. We create masks and avatars which are the best of ourselves and perform them online. Our identities become a complex interaction between the person that I am and the self that I portray on the internet. We get used to showing the world the best of ourselves and become anxious that we will be discovered as a fraud or an imposter, someone who is wearing a mask.

This makes it hard to maintain and develop a friendship when it remains purely online. Friendships online are mediated by the level of personal information which a person wants to give. It is down to an individual to bring as much of themselves as they want to the relationship. Many people do not share as much of themselves as they would in a face-to-face relationship. This can be for a number of reasons: some people just aren't all that comfortable about sharing personal information, others feel that they might be judged, and others again may just wish to present the best side of themselves to the world. I can't count the number of times I have learned something surprising about a person years after I have started working with them.

This can happen in offline, in-person relationships too, but anyone with even a small level of empathy can learn about all sorts of nuances in a person's life and world just by having a face-to-face conversation.

This impacts the form of our own identity. Our identity comes from the network that we participate in and, for many of us, this is from the work that we do. Work creates interactions and relations that contribute to the formation of who you are. When I started to work online and to form a new identity around being an online worker, my identity felt shaky. As much as I do participate in life online, I also hold something back. What I say is reflected on before I say it. Many acts are performative, intended to get people to like me, or respect me. It's easier to be effective at this online than it is offline. When I meet people offline for the first time I am often surprised by how different they are to their online persona, so I can only infer that this performativity, this selection, and this holding back is not unique to me. It means that the networks in which we operate are more surface than depth. This can make forming a solid and stable identity difficult.

Online communication lacks a lot of the context that one needs to be responsive and sensitive to what is going on in another person's life. It's impossible to know all of the contextual details of where someone is and what's going on in their environment. The other person might give you a lot of detail about the way that they are working, but since you aren't there you can't get a sense of the noise around them, the interruptions and distractions, the fact that they might be feeling a little low that day or that they've run out of coffee. It's impossible to have access to any of the appropriate cues that tell you how to react in a situation, so there is huge scope for miscommunication and misunderstanding. It's further complicated by the cultural differences between many remote workers. We don't all arrive at the screen with a shared understanding of cultural experience and idioms; sarcasm, wit, jokes, cultural references, and turns of phrase can go from light-hearted to confusing, to even insulting. I can't count the number of times I've had to look something up on Google just because I wasn't sure what a phrase used by someone else meant: phrases like "inside baseball" and "playbook" meant nothing to me until I worked online and started interacting regularly with Americans. And it's even more challenging for people who don't have English as their first language–no matter how fluent a person is, it's hard to be aware of all of the vagaries and idiosyncrasies of all of the English dialects across the world. This puts some at a disadvantage when it comes to written communication, and since written communication is the foundation of all internet relationships, it can unfairly affect them. In order to be understood, online workers have to develop an awareness of their own cultural idiosyncrasies and avoid using them, resulting in an overall flattening and homogeneity.

While the internet provides more ways for us to keep in touch with one another, when you work online with someone and strike up a friendship, it can be more difficult to maintain that friendship after the period of work ends. Say, for example, I make a friend in my co-located job, we live in the same town and go to similar places. When the work relationship ends there is still lots of scope for maintaining a close friendship. It is harder to maintain an online friendship beyond the work relationship, particularly if your relationship is grounded in a remote company. Friendships forged through online work are founded on the basis of company infrastructure. When you leave a company your access to its communication tools is revoked. Moving to unfamiliar communication tools can inhibit the easiness of a former relationship. In the move from one communication platform to another I have lost touch with a number of people I used to chat with regularly. Also, through working together you may have occasion to see one another in person, at work meetups, events, or get-togethers. When the work relationship ends because one or the other leaves the company, there is no longer any financial support to see one another. This makes it harder to maintain friendships across long distances. This doesn't mean that it isn't possible for long-lasting and meaningful friendships to form via online work; they certainly do, but they take effort to maintain and are easy to let slide. Friendships made in traditional co-located workplaces are easier to maintain because they are grounded in a physical place where you can just catch up and see one another, even when the working relationship has come to an end.

Despite this, I've found that through my years of being online, the friendships that I have formed have become richer. The people who I first chatted to online, back in the nascent days of the World Wide Web, were people who I never met, and while fun the relationships were largely superficial. In the early 2000s I started to use the internet to keep in touch with people who were already my friends, and then as blogging grew in popularity I became part of a geographically scattered community. It's really through working online though that relationships started online have become as significant as those started offline. This is because working together means that there is almost always an opportunity to physically be together, and we are given the actual space for bonds to form and strengthen and friendships to flourish.

With no opportunity and no inclination to make friends at home, I began to travel extensively for work, first to conferences but later to other events and meetups. In the past, conferences had been to me long, rather boring affairs where people presented tortuous papers on their area of philosophical research. The best part was always hanging out in the bar. It was a chance to connect with people in the same community, who were dispersed about the country. A way to connect with your wider network.

For remote workers, events like these might be the only way you connect regularly with your colleagues and peers. A conference, for me, serves as one of my main social outlets. I had spent a year at home, occasionally venturing out to a cafe to get a bit of human contact, but mostly I rattled around our small house in a tiny village. I had created a network of friends online and the only way I could see them was to meet up at tech events. For a few days, I could fly somewhere and hang out with people from all over the world, from morning to night absorbed in what I was doing, in who I was with, distant to my home.

In the tech industry, unlike the academic philosophy community, there is lots of money. Which means that there is lots of staying up all night drinking on open-bar tabs, expensive dinners, and flying to far-flung places for conferences that I probably didn't need to go to.

Going to events allowed me to connect with my growing social circles. The more I got absorbed in my work, the more alone I felt at home, the more I wanted to go away. I wanted to spend time with people who were doing the same things as me, who were part of the same community. Wherever I went, from when I arrived until I left I was part of a different world. I had cultivated a parallel life, one that only became actualised whenever we came together in person.

Back home, D felt abandoned. I rarely phoned. I was too busy to talk to him. I was too engrossed in the shiny and new. I was caught up in lengthy conversations about work, or just out to party. I wish I could say that I missed him but I was so distracted that I didn't miss him. I was fulfilling all of the needs for friendship that I wasn't fulfilling at home. I had stored up all of my socialising for these trips and once I was there I was going to extract everything from them. Not one breakfast or lunch or all-night drinking session would be wasted.

He was unhappy. He would call me and I'd only have a few minutes to talk to him before having to go. I didn't call him every day. I think he expected long conversations detailing every single thing that I was doing, like I used to do when we were first going out and we missed each other desperately and would spend hours and hours talking on the phone, but all I had was a perfunctory "hello" and an "I love you" before having to jump off to the next meeting or talk or party.

I was too immersed in what I was doing to share any of it with him. I wasn't ready for reflection. Home was a distant place, an annoyance that I had to keep checking in with, part of the life that was quiet and sedate and where not much happened.

The more I travelled the more I wanted to travel, and the more frequently I would be away from home. My trips took me farther and farther afield. First it was just places in the UK – Portsmouth, Leicester, Edinburgh – then Europe, the Netherlands, Lisbon, and then numerous trips to the US – New York, Phoenix, Savannah, San Francisco, Memphis. I transitioned during that time from straightforward freelancing to launching my own writing business. This meant I was away even more. I could expense everything to my company which made it all tax-deductible so it felt sort-of free. It was also a good opportunity for me to get new clients and meet with old ones. Not a few jobs came about over lunch in some US city. The stronger my excuse was for travelling, the more I did it, and the more D felt abandoned.

When I got home I was always exhausted, jetlagged, and uninclined to talk much about my trips. It was just business, there wasn't much to share, and this put a greater and greater wedge between us. I was creating an alternate life and it wasn't a life in the same town that he could participate in if he wanted to, it was out there, in the world, separate to everything we had at home. The distance between us continued to grow. He never stopped me from doing what I wanted to do, but I could sense that he was hurt that I had another life that he was excluded from and that he barely got to glance into. I knew that we had to fix it, to close the distance that had grown between us. But I was so absorbed in my work, so determined not to give it up, that I really wasn't sure how.

PART TWO
INSIDE THE HIGH-RISE

CHAPTER FOUR
MY SKINNER BOX
AND ME

J.G. Ballard's 1975 novel *High-Rise* tells the story of a high-rise apartment building. On the top floor live the most affluent, with the inhabitants descending in social class down to the very bottom. In addition to their apartments, the high-rise is filled with every modern convenience: a supermarket, a bank, a swimming pool, gardens, schools, high-speed elevators, hair salons, and restaurants. The high-rise meets so many needs that the inhabitants no longer have to leave; they spend all of their time inside, so that, slowly, it changes them. They are:

> [P]eople who were content with their lives in the high-rise, who felt no particular objection to an impersonal steel and concrete landscape, no qualms about the invasion of privacy by government agencies and data-processing organisations, and if anything welcomed these invisible intrusions, using them for their own purposes. These people were the first to master a new kind of late twentieth-century life. They thrived on the

rapid turnover of acquaintances, the lack of involvement
with others, and the total self-sufficiency of lives which,
needing nothing, were never disappointed.

Life in the high-rise starts to break down. Neighbours
fall out with one another, petty grievances escalate
into violence and murder. There is something about
the high-rise, about the construction of the building,
about "this hanging palace self-seeding its intrigues
and destruction," about the configuration of people
and services and conveniences, that changes the
inhabitants' behaviour, ultimately changing who
they are. The "secret logic of the high-rise" fosters
degenerative behaviour. It is the undoing of all who live
there. Tribes form, bodies pile up, society crumbles.

Ballard's tale is an extreme characterisation of
what happens at the intersection of environment,
technology, and people. The architecture of the high-
rise alters its inhabitants – it doesn't matter who they
are, everyone, rich or poor, is changed by it. *High-Rise* is
the tale of a closed environment; utopian architectural
decisions affect behaviour and cause change. When I
read the book it struck me as an allegory for how online
environments change our behaviour. The inhabitants
of the high-rise become violent, callous, and brutal. In
online environments, trolls display similar behaviour,
harassing others and making threats of violence. Is
there something about the architecture of the internet
itself and the design of internet tools that, while
utopian, has unintended consequences?

Trolling is an obvious and stark example of behaviour
that emerges through the distance afforded by the
internet. Trolls feel disinhibited because they can

attack others and have no consequences for their action. But are there other ways that the architecture is changing us? I feel that I am different to the person I was ten years ago, who is different to the person ten years before that, and not just because I am growing older. I feel cognitively different, like I am being rewired. I feel a sense of speed, that a new world is constantly being architected. This world is actively being created by someone else, but it is changing everyone who lives within it. Everything changes so fast that sometimes it's hard to get a grasp on what is happening and how it is changing us.

Interaction with technology changes how we behave, how we think, and who we are, and the major technological shift of our time is the internet. It has permeated every aspect of our lives, becoming part of the fabric of society. A thought experiment involving the sudden disappearance of the internet from our lives would lead us quickly down a dystopian path.

The internet itself is the grounding for both my experience as a remote worker and as a person within the world. It comprises many different levels, from the cables data travels along, to its deep architecture, to the code it is written with, to the websites that you and I interact with. I have found it over my life altering me to varying degrees; sometimes I am immersed fully in it, and sometimes I am able to take a step back. But I can't help thinking that I would be a different person if I wasn't so fully connected all the time. Whether that person would be worse or better is impossible to tell, but I do know that examining the nature of the high-rise is something that I can do to better understand who I am.

The idea that we are shaped and created by our environments is not a new one. In the West, Spinoza's *Ethics*, written in the seventeenth century, explores how our selves are created through interactions with our environment; in Hegel's *Phenomenology of Spirit* self-consciousness emerges through a dialectical struggle. In the East, the idea of a relational self is much older, and goes back as far as Confucius.

In the twentieth century, the concept of relationality was given scientific validity with the discovery of neuroplasticity, the discovery that, without medical intervention, our brains can be reprogrammed. Previously, neuroscience held that the brain develops during early childhood only to become stabilised and "static." But research conducted in the late twentieth century has revealed that brain plasticity continues throughout adulthood. This means that the brain can change in function and structure without any medical intervention. It's affected by the things that we do every day, the environment we engage with, the world we live in. Through repetition we can form habits, and the more we persist with a habit the more entrenched it becomes. We can, however, change our habits through sheer persistence that rewires our neural pathways. This is captured in the neuroscientific adage, "the neurons that fire together wire together."

The roots of research into neuroplasticity can be found in studies into whether the environment has any impact on an animal brain. Researchers studied rodents in enriched and unenriched environments: those in the enriched environment had toys and stimulation, those in an unenriched environment had none. Autopsies on the animals showed significant differences in the two groups. Those kept in an enriched environment had more cellular connections, a larger cortex, and had formed new brain cells in the hippocampus.

More recently, there has been research that explores how the human brain changes. In the UK, London taxi drivers must learn "The Knowledge," the complex network of streets and landmarks of London, in order to operate a black cab. A 2000 study, and a later follow-up, found striking changes in those taxi drivers who, after three to four years' training, successfully acquire The Knowledge: they have greater volume of grey matter in the posterior hippocampus (the area where brain processing takes place), and less in the anterior hippocampus relative to non-taxi drivers; they also display greater memory for London-based information and poorer memory and learning in other memory tasks. While they have increased their brain capacity in one area they have limited it in others.

The evidence for neuroplasticity is piling up. A 2007 study found that when monkeys are trained to use tools it changes their cortical motor neurons so that they act as though pliers are their fingers. In 2010, a study at the University of Goldsmiths observed that just thirty minutes of voluntary control of brain rhythms creates a lasting shift in brain function.

Neuroplasticity has powerful implications for those who have suffered brain injuries. If the brain is not static it means that there are possibilities for treatment and brain therapies that would have previously been unthinkable. Michael Merzenich, for example, one of the pioneers of neuroplasticity research, has used his findings to remediate individuals with speech, reading, and language deficits.

The concepts of neuroplasticity are also being used, unsurprisingly, by the self-help industry, which sees it as proof that through the power of positive mental thinking or mindfulness, you can change yourself for the better. You can form habits, good habits, that help you in the endless quest for self-improvement. Perhaps this might be the case, especially if the person is able to cloister themselves away from the rest of the world. However, the brain is not in a vat. It is embodied in a world with power structures that constantly act on it–society's norms, normative ways of thinking, the ideological stories told by governments, institutions, society, and media. Any individual action to affect the brain's functioning will always be in tension with the wider environment.

In his book *The Shallows: How the Internet Is Changing the Way We Think, Read and Remember*, Nicholas Carr draws the relation between neuroplasticity and the internet, arguing that the internet is not just changing our lives but is changing the way we think. He uses as his starting point the theories of Marshall McLuhan, who coined the famous phrase "the medium is the message." When a new media form is introduced, what matters is not the content (whether that be the written word, movies, music, spoken word, or other forms of communication) but the medium itself. While we are distracted and dazzled by the content, the medium is subtly and anonymously shifting the way that we see the world, creating new patterns in our brains and changing our behaviour. McLuhan writes, "The effects of technology do not occur at the level of opinions or concepts, but alter sense ratios or patterns of perception steadily and without any resistance."

McLuhan's insight is that technologies change us, and they do so anonymously. The last thirty years have seen the rise of a technology that has transformed how we think and relate to one another. We have gone from one-to-one interactions to a globally networked society in which communications happen simultaneously, everyone is connected to everyone else, and information moves practically at the speed of thought.

Like the inventors of most technologies, the inventors of the internet cannot have been aware of the consequences that it would have for society. Just like Anthony Royal, the architect of Ballard's high-rise, there's no doubt that they were utopian in their aims, seeing this new network as a way to provide universal access to information, but they cannot have seen how the ontological fabric of the internet would come to impact upon our world.

The predecessor to the internet was ARPANET, a packet-switched network which was based on the concepts of multiple researchers who converged at the US Department of Defense's Advanced Research Project Agency (DARPA) and other institutes. In the early 1960s multiple theories came together to provide its basis: the concept of the Galactic Network; Paul Baran's conceptual model for a distributed network that would preserve communication channels in the event of a Soviet attack; and packet-switching. Packet-switching is a way of distributing data on a network: a piece of data is broken down into small blocks, called "packets," and these blocks of data are transmitted in pieces along the network and reassembled at the receiving end. Traditional analogue telephony uses a process called circuit-switching in which two nodes require the opening of a continuous dedicated communications channel. On a packet-switched network, packets are passed along the network simultaneously, allowing for multiple communications to happen at the same time. The network was further refined in the 1970s, with the introduction of the Internet Protocol Suite, or TCP/IP, which specifies how data should be packetised, addressed, transmitted, routed, and received. TCP/IP shifts responsibility for the reliability of data from the network to the hosts, allowing the shift from separate closed networks to multiple connected networks, what was then called "internetworking."

It wasn't just scientists and engineers who were catching on to the power of distributed networks. In 1974, the author Arthur C. Clarke was asked in an interview what the world would be like in 2001. Clarke said that everyone would have a console in their home, that a person would be able to "talk to his friendly local computer and get all the information he needs for his everyday life, like his bank statements, his theatre reservations – all the information you need in the course of living in a complex modern society." He goes on:

[I]t will also enrich our society, because it will make it possible to live really, anywhere we like. Any businessman or executive can live anywhere on earth and still do his business on a device like this, and this is a wonderful thing, it means we won't have to be stuck in cities, we can live out in the country or wherever we please, and still carry on complete interaction with other human beings as well with other computers.

Over the thirty years following the establishment of ARPANET, other distributed networks sprang up around the world as different research institutions, military establishments, and, later, commercial bodies saw the potential of distributed, packet-switched networks. Many of these networks provided the basis for today's internet.

It was in 1989 that the internet took the form that we're most familiar with – the World Wide Web. Invented by English computer scientist Tim Berners-Lee, the World Wide Web uses a domain-name system to replace IP addresses – locations on the network have memorisable names instead of a series of numbers. It uses a non-linear text system called hypertext (also referred to as hypermedia). Hyperlinks are used to navigate the web, and the list of links on each page is more important than the content. Importantly, it is a network on which anyone can create resources. By the end of 1990, Berners-Lee had created all of the tools for the web: Hypertext Markup Language (HTML), Hypertext Transfer Protocol (HTTP), the first web server, the first web browser (called WorldWideWeb), and the first web page.

Berners-Lee made the web freely available, with no patent or royalties due. In 1994, he founded the World Wide Web consortium (W3C), which is the body that creates standards and recommendations to improve the quality of the web. By setting standards, W3C ensures that the web is displayed consistently across web browsers. One of its founding principles is that any technologies it recommends must be royalty-free so that they can be easily adopted by anyone.

The World Wide Web became the gateway to the

internet. Anyone with a computer, a web browser, and a connection could get online and access it. Between 1995 and 2016, the percentage of the world population with access to the internet grew from 0.4% to 50.1%. The internet and the World Wide Web are the technologies that have defined my generation. I can recall a time without the internet access, but it seems like a really, really long time ago, when the PC was used to play *Lemmings*, I was restricted in my TV-viewing to what was on right at that moment, and I had to call people on the telephone to make arrangements to meet up with them.

The internet is having a profound impact on the way that we think, and the more that we use it the more marked the impact. We talk about innovative disruption when talking about internet companies, but the greatest disruption is the one carried out by the internet itself. It hasn't just disrupted an industry but has disrupted our being in the world as we take on the ontological contours of the internet itself. By looking beyond the surface of the internet and interrogating our own experiences, we can see some of the ways that the internet has changed us.

We have become packet-switched. When we communicate online we interact with information and with each other in ways that mirror the way that the internet itself operates. We are always dealing with fragments. In my daily life I deal with multiple communications at the same time: numerous private messages, multiple streams of chat, fragments that come from social media. It's rare for my focus to be on just one conversation. This means that, just like the internet, I receive packets of data and my brain reassembles them into the correct order.

Alongside this neural packet-switching, we are affected by the instantaneous nature of online communication. The World Wide Web is built using a protocol called Hypertext Transfer Protocol (HTTP). HTTP is a request-response protocol: a client (a browser or application) submits a request to a server (where all of the data is stored). The server returns a response with all of the information that the client has requested. In modern computing, this happens in a matter of seconds and is happening at increasingly fast rates. HTTP is the structure of the web, its ontological underpinning. In the internet medium the request and response are part of the same thing. When a request is made a response is automatically formed. Our mode of communication becomes one that is automated and instantaneous.

Beyond communication, the World Wide Web changes how we deal with information. There is a vast difference between reading a book and reading a piece of hypertext. When I read a book and am deeply immersed in it all sorts of unexpected connections swim to the surface of my mind, things that seem completely unrelated, comparisons are drawn between things that are unexpected. These sorts of connections are places where the imagination takes hold and we can form novel and unforeseen ideas. Text on the internet is hyperlinked: the connections have already been made. All I have to do is click on a link and I'll be taken to something that is related to that text. These are connections between things that are obviously related, that are on the surface of the text. I never click on a hyperlink to be taken to something that completely baffles me. These connections don't spring from our own memories and experiences but are given to us by the author of the text. Over time, as we become more and more reliant on hypertext to do this work for us, we lose our intuitive ability to create connections ourselves. Hypertext keeps us on the surface, gliding along what is given, rather than drawing on our own knowledge and experience to make connections ourselves.

Fragmented, instantaneous communication, high cognitive load as we parse together information, and a surface that contains everything; these are some of the conditions of working on the internet. It is fast, fragmented, and encourages all of our brains to follow similar paths.

Michael Merzenich writes:

> There is absolutely no question that modern search engines and cross-referenced websites have powerfully enabled research and communication efficiencies. There is also absolutely no question that our brains are engaged less directly and more shallowly in the synthesis of information, when we use research strategies that are all about "efficiency", "secondary (and out-of-context) referencing", and "once over, lightly" ... THEIR HEAVY USE HAS NEUROLOGICAL CONSEQUENCES. No one yet knows exactly what those consequences are. (His capitalisation)

These changes could not have been predicted by the architects of the internet and the World Wide Web. Those designing and building the web today, however, are more aware of how the tools they build can shape the way that we behave. It's common to hear designers talk about how they "design experiences," creating an experience for a user, one that is easy, or pleasurable, that keeps them coming back to the platform. User-experience design make use of concepts such as "user stories" or "user journeys" – pathways that the designer wants you to take as you navigate through a website. This could be positioning a button in a certain place so that your eye is drawn to it, using language that encourages you to take an action, using colour to draw your attention to a specific website element, or ordering menu items in such a way that the ones the designer wants you to click on are most prominent on the menu.

Many of these techniques and tropes are underpinned by theories of behaviour design, an academic discipline that underlies many of the most popular and common approaches to web design.

Behaviour design involves designing software in such a way as to affect people's behaviour. The goal is to persuade users to take an action. The father of behaviour design is B.J. Fogg, who runs the Persuasive Tech Lab at Stanford University. Fogg's behaviour model proposes that three things need to converge for a person to take action: they have to want to do it (motivation), be able to do it (ability), and be prompted to do it (trigger). In his article "A Behaviour Model for Persuasive Design," he gives the basic example of a website owner who wants to get people to sign up to their mailing list. The target behaviour is typing in an email address. While this action is easy for people to do, what fluctuates is a user's motivation to do it. Some people will be really interested in the mailing-list content, others may not care. With the right trigger, those who are motivated will sign up.

There is a relationship between motivation and ability: if a person is highly motivated they will take an action even if it is difficult, and an unmotivated person who encounters an easy action is more likely to take it just because they can. In behaviour design, these levels of motivation and ability can be manipulated, increasing their motivation (offering an incentive, for example) or increasing their ability to take an action by making it very easy.

Fogg identifies three motivators. The first is pleasure/pain, a primitive and immediate response, which Fogg acknowledges may not be the best model for designing behaviour. The second is hope/fear, a motivator characterised by anticipation: we hope that something good will happen and we fear that something bad will happen. People sign up to a dating website in the hope that they will meet someone, and they update their computer for fear of viruses. The final motivator is social acceptance/rejection. We can see this motivator at play in social media like Facebook, Twitter, and Instagram, where we do things in order to be accepted (get Likes or other forms of approval), and because we fear social rejection (censoring those things that we think will make us look bad).

The second element, ability, is concerned with a person's ability to take an action. Usually it concerns the ability that someone actually has at that moment, rather than their potential ability. Potential ability can only be harnessed through teaching, teaching requires effort, which lowers their capacity to do something easily. Instead, to increase someone's capacity, actions need to be simple. The designer needs to remove any barriers that get in the way of someone's ability to take an action. Amazon, for example, always makes it easy for you to buy by saving your payment details and offering a 1-Click Purchase.

In general, Fogg advocates always trying to make a task simpler rather than trying to increase a person's motivation because "making an action easier takes less work than increasing a person's motivation." This is why so many of the actions we encounter on the internet are so easy to take; whether it's a Like, or a Share, or a purchase, they are as frictionless as possible.

While playing with various levels of motivation and ability, behaviour designers need to implement triggers. Triggers make you take an action. Many of these triggers are intensely annoying: no one enjoys being harassed by a pop-up advertisement to enter your email address for a mailing list, or having to turn off a video that loads on a page. Triggers need to be well-designed, well-timed, and well-positioned. When a trigger happens at the correct convergence of motivation and ability, the action will take place. The trigger could be positioned to get you to make a purchase, donate to a cause, crowdfund a product, leave a comment, etc. On Facebook I receive a trigger message that tells me I've been tagged in a photograph: clicking on the message to view the photograph is easy and I am motivated to do so as I want to see the photograph. In contrast to television, the internet allows us to make immediate responses to triggers, often in a split-second. And because these responses are low-friction and don't require much effort, we're more likely to make them. Some require no action at all: take for example Netflix's "playing next" feature. You reach the end of an episode of your favourite television show and the next episode is queued to play. This trigger converges with your motivation to want to watch the next episode, and your ability to take action – no ability is even required since it is easier to just let it keep playing.

One of B.J. Fogg's maxims is "put hot triggers in the path of motivated people." A hot trigger is something that you can click that immediately takes action – "click this link," "view this image," "buy this item," "enter your information." These are things that someone who is motivated can do so easily that there's no reason for them not to do it. A hot trigger is an action you can take immediately, versus a cold trigger, such as a radio advert that you hear while driving suggesting you buy something. It's cold because you can't take action immediately.

Internet platforms are not neutral spaces. They are spaces where designers are constantly trying to manipulate behaviour. They make actions frictionless to increase the chances of us taking them. Fogg's behavioural design model has permeated far and wide; there are many Silicon Valley entrepreneurs who have taken his course and used the theory to spin up hugely successful startups.

One of those students is Nir Eyal, who published the book *Hooked: How to Build Habit-Forming Products*. Even the title of the book has echoes of addiction, promising a guide on how to make your users addicted to your product. In an article about Instagram's addictive qualities, he unproblematically refers to one user as an addict, and goes on to laud this as the platform's success. It's addiction to a platform that so many designers are aiming for.

Unlike Fogg, whose theory is built on the premise of external triggers–i.e. triggers on a website–Eyal builds his theory on the idea of internal triggers. It's not enough to just "put hot triggers in the path of motivated people"; a designer needs to motivate people to visit a website or platform without any external prompting. The product needs to become part of their daily habits, part of the user's routine, so they visit it without even thinking about it. Waiting in a queue at a shop or standing at a bus stop becomes an automatic cue to check Twitter, look at your Instagram, check what your friends are up to on Facebook.

For Eyal, the product-maker needs to manufacture desire, and they do this by creating hooks. Like Fogg's behavioural model, this is based on a convergence of motivation, action, and triggers. Unlike Fogg's model, there is a distinction between internal and external triggers. The product-maker first presents the user with external triggers–an email, an icon, an alert, or a link. The user takes the action and when they start to do it frequently they create associations with the external triggers, attaching them to specific emotions: their need for social approval, for example, or their desire to be entertained. It doesn't take long for internal triggers to form so the user takes action without even thinking about it.

In order to create desire, feedback loops cannot be predictable. Therefore, once the user has taken action, they are given a variable reward. If I know that I will always get Likes on my photo and I always know the amount of Likes, there's no reason for me to return to the platform to check. Instead, I feel anticipation for a possible reward–will I receive it or not? I have to keep looking for it. A checking-in platform like Swarm provides variable rewards in the form of points. I check in at a location and I get points, but I get more points if I post photos or check in with friends. I'm never 100% sure how many points I will receive, but when I get them I can see myself on the leaderboard and see how much better I am at checking in than my friends. Variable rewards keep people coming back; exactly the same theory that slot machines are based on.

Finally, the user has to do a bit of work. They invest back into the platform and their investment makes them further implicated in it. As Eyal says, "now that the user's brain is swimming in dopamine from the anticipation of reward in the previous phase, it's time to pay some bills." This rarely involves paying for anything, but usually takes the form of time investment or social capital–adding friends, Liking other things, leaving comments, filling in your profile or forms, all of which create a further commitment to the platform.

Unlike traditional advertising, which creates a feedback loop linking a product to a reward–"this car will get you where you want to go faster," "buy this hoover to make your wife happy," "milk makes you healthy"–hooks create habits that create behaviours without the need for any external triggers. The user is altered over time by interacting with the product, acting in ways the product determines.

The theoretical forefather of both Fogg's behavioural model and Eyal's theory of hooks is B.F. Skinner.

Skinner was a behavioural psychologist, a branch of psychology that was founded around 1913 but that fell out of favour in the mid-1950s. Behaviouralism is a theory which says that all behaviours are acquired through conditioning–we are changed and shaped by our relationship with our external environment. Skinner ran a series of experiments using an operant-conditioning chamber, or Skinner box. The subject in the box would be trained to respond to stimuli, and when they correctly performed a behaviour they would receive a reward. A lab rat, for example, pulls a lever and is rewarded with food. Through continual positive reinforcement, behaviour is changed. Skinner discovered the subjects respond better to a variable-rate schedule of rewards, rather than a continuous rate. Over time, the subject is conditioned to act in predictable ways.

For product designers we are all lab rats acting predictably in response to our reward loops. Our behaviour is changed. We are altered. We spend our days engaging with technologies whose explicit goal is to alter our behaviour. These interactions become internalised so that how we act changes over time. Who we are and how we act are two expressions of the same thing, and in altering one we alter the other. These are being quietly influenced by a small group of men in Silicon Valley.

Both Fogg and Eyal acknowledge that their behavioural models aren't necessarily good. In an article in the *Economist*'s 1843 magazine a somewhat circumspect Fogg is quoted as saying: "I look at some of my former students and I wonder if they're really trying to make the world better, or just make money.

What I always wanted to do was un-enslave people from technology." Eyal is more nonchalant: he admits that habit-forming design can be used for good or bad, but says that companies should be using the theory to make their products, and places responsibility on users "to understand the mechanics of behaviour engineering to protect themselves from unwanted manipulation."

One of the main drivers of revenue on the internet is advertising, and in order to advertise to people you need to keep them on your platform, keep their attention, keep them addicted. In Q1 2016, Facebook's advertising revenue increased 57% to $5.3 billion, and in Q2 it reached $6.24 billion, far exceeding the $5.8 billion predicted. At the end of Q1 2016, 97% of Facebook's overall revenue came from advertising. This mean that Facebook wants to keep you on the platform. Its means for doing that is commanding your attention. Without attention, Facebook users will just go elsewhere.

There is beginning to be a backlash against this approach to design. Tristan Harris, for example, is a former "Product Philosopher" at Google, and the founder of Time Well Spent. He's also an alumnus of B.J. Fogg's Persuasive Technology Lab. In an article on *Medium*, Harris compares the activity of product designers to the activity of magicians – both look for blind spots that they can exploit, using your psychological vulnerabilities to make you consciously or unconsciously take an action.

Product designers make choices that get you to make decisions that meet the goals of their product. They wouldn't be very good product designers if they didn't. Harris goes through some of the ways that product designers are able to hijack users. These include:

controlling menus, Fear of Missing Out (FOMO), exploiting our need for social approval, exploiting our sense of social obligation and reciprocity, infinite feeds and autoplay, instant interruption, making choices inconvenient, and exploiting people's inability to forecast the results of their choices (would anyone go to Facebook to check out an image if they knew they would spend the next twenty minutes on the platform?).

Harris also highlights some of the effects of having multiple interactive technologies in your smartphone. Because intermittent variable rewards are so powerful, and because each application offers its own intermittent variable rewards, there is an accrual of rewards so that almost every time you go to your phone you can expect a new reward, whether it's a Like, an email, a tweet, a text message or whatever. This wasn't intended by any individual software manufacturer, but is due to there being so many of them. As a result, people use their smartphones a lot. In a study by Nottingham Trent University, participants used their phones for about five hours a day and checked their phone about eighty-five times.

In addition to all the time spent checking and re-checking our phones or applications, these interruptions cause stress and frustration. Research by Gloria Mark at the University of California found that "after only 20 minutes of interrupted performance people reported significantly higher stress, frustration, workload, effort, and pressure." A constant chain of interruptions throughout your day serves to make your working life more stressful.

Some applications try to solve this by allowing the

user to adjust or turn off notifications. Skype has multiple status modes, including Do Not Disturb, which allow the user to indicate to others that they're not available. In December 2015, Slack introduced a Do Not Disturb mode, making it possible for a user to turn of notifications and focus on what they need to do.

However, much of the responsibility for controlling the flow of information falls back upon the individual. While it may be a goal of Tristan Harris and others to introduce ethics into product design, I am more skeptical. Companies need to keep users on their platform, keep their attention, manipulate their decisions. Otherwise, who would they advertise to? We live in a world where people are expected to make informed decisions themselves, and yet there is a shortage of information around that would help people make those decisions. I only am aware of behaviour design because I'm writing about how the internet affects us. I knew little of it before starting this journey.

There are things that we can all do though. Some people give up their smartphones, instead having "dumb phones," mobile phones that don't have any access to the internet. For me, that's one step too far. While a smartphone can be a pain in the ass, it comes with many things that enhance my life. Having a GPS in my pocket helps me to get where I want; I have thoroughly documented the early life of my kid with my always-available camera; I have an immense archive of music at my fingertips; I can always find the best restaurant in town; I can easily plan travel, convert currency, order food or taxis, or listen to music and podcasts. Why would I give all of that up?

That doesn't mean that it doesn't also diminish

my quality of life. I find that if I have social media on my smartphone I am constantly reaching for it, both consciously and unconsciously. When I catch myself doing so I am reminded of the pull they have on me. So I deleted Facebook and Twitter from my phone. When that didn't go far enough and I found myself checking them in the native web browser, I deactivated Facebook and changed my Twitter password to be so long and complicated that it made it very difficult for me to enter it. These steps are not cure-alls, but I've found that the fewer communication media I have on my phone, the less of a pull it has on me.

It's at this intersection between the deep architecture of the internet and the surface design of internet platforms that I find myself changed. The ontology of the web grounds a fragmentary self that is defined by the speed of its interactions and its relations within a shallow, ever-changing network. These are rather shaky grounds for subjectivity. This self is characterised by the speed at which it communicates and the surface connections along which it passes. It is constantly triggered to take simple actions, ever-rewarded by short-lived moments of pleasure. Across distributed networks it performs for social approval, anxiously searching for acceptance and only pausing for a moment when it gets it. This self is constantly being defined by a high number of relationships which it never has adequate knowledge of.

It is within this context, at the end of the twentieth century and the beginning of the twenty-first century, that we are told that we have a high level of individual liberty and freedom. These changes brought in by technology have been shaped and harnessed by other forms of power. If we accept that our brains can be changed by technology, and neuroscience tells us that this is true, then our brains can be changed by other interactions, by the invisible power structures and flows that make up society. By the shaping that comes from education and through constant exposure to the media. Mirroring the growth and spread of the internet other global changes and flows of power have had just as much impact. We are altered by the very visible changes in technology, but that has happened alongside a revolution in politics and economics. These political changes have developed along a similar timeline to the internet, and its ideas of freedom and individualism have become dominant principles on the internet. Just as much as we have been shaped by the internet, we have been shaped by the ideology of neoliberalism and one of its familiar expressions, the entrepreneur.

CHAPTER FIVE
WE ARE ALL ENTREPRENEURS NOW

The website Founders and Funders has an infographic. Its title is "Everyone will become an entrepreneur: How the world will replace employees with entrepreneurs." A cheerful infographic outlines how "most jobs" will be done by independent contractors, which is great for businesses who will save 30% on employee costs. Businesses can "pay as you go," just bringing in contractors when they need them. This means no supervision costs and no money wasted on benefits: this last point is made with the icon of a baby with a red line through it.

The predictions don't stop there: companies will outsource to countries such as India, Russia, or China, where labour is cheaper; manual jobs will be done by robots; corporations will expect employees to act as "intrapreneurs"; creative jobs will be carried out by people "on the side"; and small business owners will be expected to become "jacks of all trades." In text surrounding the infographic, the author is effusive about how companies don't want traditional employees anymore, how the only way to get hired is to be an entrepreneur, to have the ability to "create value out of nothing." The author says: "For those who do not like this whole thing, this news is actually good. The world is becoming better, the only thing is you also have to change with it – and the best way is to become an entrepreneur."

The infographic captures what startup culture thinks about both work and reality: that we live in a world where everything comes down to individual responsibility, that we are all expected to take risks, and a company or business has no duty of care to those who are responsible for its success. In graphical format, it captures how startup and business culture present the logic of the market as reality.

I came across the infographic at a turning point in my own career. It was 2013. I had gone from straightforward freelancing to running my own successful copywriting business and was thinking about whether to expand it or whether to do something else. I was already subcontracting work to a number of other writers and, with energy and commitment, I could establish it as a proper writing agency and grow it to have employees, with me as the CEO. It was tempting. Many of the people I had met through my career working online had established businesses which were growing rapidly. They were becoming successful in a way that was very easy to understand; after all, who doesn't want to found and run a successful startup?

When I launched my business in 2011, I received an email from an entrepreneur congratulating me on launching my startup. Startup? I recall being both flattered and confused. I didn't feel like I had a startup–it was really just a shopfront for me to sell my writing skills. I wasn't incorporated and I hadn't yet started subcontracting work. It was just me. It didn't feel very startupy. But maybe it was, maybe I was an entrepreneur.

There were certainly aspects of it that were entrepreneurial: I was taking a risk, giving up a relatively stable freelance contract and betting on my own ability to generate business; I had seen a gap in the market that I knew I could fill; I knew I needed to be a jack of all trades, understanding everything from my own craft of writing to elements of software development, user experience, and quality assurance, as well as bookkeeping, client relations, and generating business.

I may not have felt like an entrepreneur but it was flattering that other people saw the entrepreneurial in me. After all, all around me I saw successful entrepreneurs, people striving to be successful entrepreneurs, articles about how to be entrepreneurial. I worked on the internet, a medium that empowers go-getters and upstarts to take control of their lives, be their own bosses, and break free from the rat race. It all resonated deeply with me.

And yet, when it came to taking the next step, to going from being a small outfit to a proper company, I was on the fence. I didn't know if I wanted the responsibility of people working for me, I didn't know if I wanted to spend my days and my nights focused on running a business, I didn't know if I wanted to be an entrepreneurial role model, a business-woman. I was excited by the fast pace and opportunities of being an entrepreneur on the internet, but when I saw that

Founders and Funders infographic something didn't sit right; I knew that there was something skewed around all this talk about entrepreneurship. I had a feeling that the rhetoric around entrepreneurialism hid something else, but I couldn't quite put my finger on it. The longer I've worked on the internet, the longer I've engaged with the world of startups and entrepreneurs, the more that feeling has grown. Everything I've read pushes entrepreneurialism as aspirational, but the more I've become immersed in it the more I see it everywhere and the more skeptical I have become.

To understand what made me so unsettled about entrepreneurialism I needed to first understand what an entrepreneur is, where the ideas come from, what makes the entrepreneur such a revered figure today, and, perhaps just as importantly, the dominant neoliberal ideology of which it is an expression. Neoliberalism, as much as technology, has created conditions that have changed our working lives, both through the casualisation of work and through the creation of effective remote workers. If you're already tired of discussions of neoliberalism and entrepreneurialism that you have read elsewhere, you may wish to skip over the next few pages. Some basic understanding of it is necessary, however, to understand where we are today and how we got here.

The basic dictionary definition of an entrepreneur is someone who sets up a business or businesses. It is a loanword, taken from the French *entreprendre*, which means to undertake. This direct translation presented a problem for early translators of writing about entrepreneurs. Jean-Baptiste Say, in his *A Treatise on Political Economy* (1821), was one of the first to write on the subject. The translators of the 1855 English edition note the challenge of translating the word *entrepreneur*, "the corresponding word, undertaker, being already appropriated to a limited sense." They define this person as someone who "takes upon himself the immediate responsibility, risk, and conduct of a concern of industry, whether upon his own or a borrowed capital." The word they settle on is "adventurer."

The entrepreneur-adventurer is someone who has a stake in a business undertaking, whether their own funds or someone else's. They participate in any industry, from agriculture, to mining, to weaving. They plan and invest in the future growth of their business: if they rear animals they purchase more land or improve what they have, merchants buy and sell in increasing quantities. They speculate on future public demand and invest capital in meeting those demands. Sometimes it pays off, sometimes it doesn't. They broker deals amongst vendors and purchasers. Say's definition of an entrepreneur isn't just someone who runs a big business. He notes that artisans, who make profits not wages, can also be entrepreneurs. Watchmakers who transform metal into watch chains and sell them at a profit are entrepreneurs, as is the flax-spinner who buys some flax, spins it, and turns it into profit.

Say wasn't the first to use "entrepreneur" in a business sense. Seventy years prior to Say, the Irish-French economist Richard Cantillon used the word "entrepreneur" to describe those people who buy from market towns and transport those goods to larger towns for sale. They also bring goods to market towns to trade with villagers. Cantillon noted that entrepreneurs are bearers of risk: some get rich, others become bankrupt. His theory of entrepreneurship is one in which entrepreneurs are directors of resources. As with Say, for Cantillon anyone who takes on risk by speculating on demand – whether they be a draper, shopkeeper, or farmer – is an entrepreneur.

The term "entrepreneur" was broadened even further by John Stuart Mill who, in his *Principles of Political Economy* (1848), briefly mentions an entrepreneur as a person who assumes both the risk and the management of a business.

The ideas of these three thinkers contain some of the nascent qualities we today associate with entrepreneurship: someone who takes risks, who has a stake in a business, who speculates on future demand, who moves around resources. However, it's not until a hundred and fifty years later that the economic conditions become such that the entrepreneur is the hero of the day.

In the late nineteenth and early twentieth century a new group of economists took up the term "entrepreneur" and defined entrepreneurs in a way we are more familiar with today. The Austrian economist Joseph Schumpeter is closely associated with entrepreneurship with his theory of "creative destruction" (1943). Creative destruction involves bringing new innovations to the market and in doing so destroying old ways of doing things. For Schumpeter this is one of the essential aspects of capitalism: the old is repeatedly destroyed in the creation of the new. This results in the constant churn of jobs and the continuous demand on people to change and adapt to new conditions. The entrepreneur is an agent of creative destruction.

Today "creative destruction" is usually called "innovative disruption" and you don't have to look far to see how new technologies disrupt old ways of doing things. Uber (taxis), Deliveroo (food delivery), Netflix (film and television), Spotify (music consumption), Airbnb (holiday rentals), and Amazon (books, and later consumer goods), are all companies that have had a huge impact, obliterating, or at least placing under threat, previously stable industries.

Schumpeter was part of a vanguard that started sketching out ideas that we take for reality today. For him, the entrepreneur is a hero, a pioneer, the archetype that we should all aspire to. The entrepreneur is:

> exceptionally free of tradition, he is the true severer of all connections, equally alien to the system of super-individual values as to the class he comes from and the class he is entering: he is the special trailblazer of the modern human and the capitalist, the individualist way of life, the sober habit of thought, the utilitarian philosophy - the first brain capable of reducing beefsteak and ideal to one common denominator.

Other economists and theorists have described different aspects of what it means to be entrepreneurial. For Frank Knight, an entrepreneur is someone who can deal with a high level of uncertainty. Knight distinguishes between risk, which is knowable and therefore insurable, and uncertainty, which is not. Entrepreneurs take chances in certain markets and make decisions that they cannot know the outcome of. For this reason, they assume a high level of responsibility, both for themselves and their businesses. Always built into this uncertainty is the possibility of failure, and many entrepreneurs do fail. In order to mitigate against this chance of failure the entrepreneur needs to be tolerant of uncertainty and be self-confident. They also need to be good at making judgements in conditions that are always changing. Without these faculties, the entrepreneur becomes frozen in a fast-moving market, unable to be entrepreneurial.

Israel Kirzner has identified "alertness" as another entrepreneurial feature. This is the ability to recognise opportunities when they arise and to take advantage of them. Alertness is not something that can be learned but is an innate quality that sets the entrepreneur apart from other people. They can identify which ideas they should bank on, find gaps in the market to exploit. They are quicker than others to exploit these opportunities and know when it is time to make their exit.

More recently, Edward Lazear has argued that entrepreneurs succeed because they are generalists. They do not excel in any one skill but are competent in many. A study of Stanford University Business School graduates showed that entrepreneurs invest their time in learning multiple skills, studying a more varied curriculum, taking on many different roles and gaining some experience in them before starting a business. This knowledge of a broad group of areas means that they are good at hiring people.

An entrepreneur possesses the ability to spot talent and to bring together people with a variety of skills that complement one another. People who just want to specialise in one skill do better as employees, with a preference for job security and an opportunity to focus on their skills and talents in one specific area. A 2013 study expanded on this jack-of-all-trades theory. It looked not just at the diverse set of skills that an entrepreneur must have but at individuals' social networks. It found that entrepreneurs have a broad social network, with lots of contacts.

There are numerous other qualities associated with the entrepreneur, including an ability to unite people around a single vision, to delegate well and find smart people for the right jobs, to be visionary and single-minded. The entrepreneur is someone who expresses their creativity through economic means. They are able to identify opportunities in the market and make the most of those opportunities to build a business, to make profits, to launch products, to individuate. It's possible to achieve a deep level of satisfaction and self-fulfilment within this. When I reflect on my own career, those points at which I have acted entrepreneurially have been times when I have seen an opportunity and made the most of it, usually to my own advantage. These moments are marked by short-term feelings of excitement, commitment, and, if they are successful, personal fulfilment. This contemporary world of which we are part, with its constant change and momentum, is a place where there are more opportunities for this than ever.

In more abstract terms, entrepreneurs are individuals who adapt themselves to market forces. Over the past forty years the business-person and the entrepreneur have become society's heroes. This ascent of the entrepreneur to the position of role model has been naturalised but, like the succession of any archetype, it is a symptom of an ideology. As the internet has become part of the fabric of our lives, so, invisibly, has the ideology of neoliberalism, which finds its most perfect expression in the entrepreneur. There can be no free market without the entrepreneur. The entrepreneur keeps the free market dynamic and is the engine of cultural and economic change; the entrepreneur is the agent of neoliberalism.

The philosophy of liberalism dates back to the late eighteenth century, later splitting as a theoretical framework into economic liberalism and political liberalism. The basis of economic liberalism is the idea of competition: in order to create a market there must be at least three individuals–two competing sellers and one buyer. The presence of the buyer brings prices down. This is the simplest type of "market force"–a nonhuman force that is created simply by the presence of three people. When there are billions of people with competing drives, needs, and wills, market forces become increasingly powerful and complex.

Liberals see the market as essentially good and want to see it propagated widely. Anyone should be able to participate in the market. They dislike any sort of self-sufficiency and anything, such as trade barriers, that prevents openness or that creates barriers to individuals participating in the market. They believe that aspects of society such as the distribution of wealth should be carried out by the market, and are hostile to any sort of interference in that market, whether by the church or the state. For the liberal, all goods and services are produced by the market. It is within this context that the entrepreneur emerges.

Economic liberalism grew in prominence in the Enlightenment as an alternative to feudalism and mercantilism. The first thinker of economic liberalism was Adam Smith, whose theories are still influential today. Smith advocated minimum state intervention in markets. He proposed the theory of the "invisible hand": that when an individual acts to benefit themselves they may benefit everyone more than if they just worked to benefit society. In a liberal society, everyone is working for themselves and society as a whole benefits. Allowing individuals to act unfettered leads to a balanced society. The role of the state is to ensure justice and equality for all, enhancing the individual's ability to act without interference.

In the battle with archaic feudal systems that insisted on the rights of the aristocracy and the divine right of kings to run economies in their own personal interests, liberalism eventually won out, and by the end of the nineteenth century many Western economies were run along liberal terms.

The ideas of Adam Smith and other liberal theorists such as John Locke and John Stuart Mill underpin today's neoliberal ideals, which are an acceleration of the project begun with liberalism. Neoliberalism is a mixed bag of theories that have in common a commitment to the free market, individualism, and entrepreneurial ideals. Unlike liberalism, its theoretical forefather, neoliberalism seeks to constantly expand the market into all areas of life.

It is not concerned just with goods and services but constantly seeking new areas for marketisation.

One of the forefathers of neoliberalism is Friedrich von Hayek, whose hugely popular *The Road to Serfdom* has become one of the free market's defining texts. Written in the aftermath of World War Two, it argues that National Socialism was only able to emerge because of the central planning and market interference carried out by socialists. Although well-intentioned, socialism always interferes with individual freedom. It centralises power in one body rather than distributing it across individuals, and, in order to maintain its conception of equality, leads to coercion, and dictatorship is an inevitable result. This means that socialism, Keynsianism, central planning, and any interventionist economics will always fail and the result, for the majority, is serfdom. In Nazi Germany, socialism, with its focus on central planning, provided the grounds on which fascism was able to emerge. Competition, on the other hand, is a force that does not require coercion but that operates by itself, free from intervention. Hayek argues that no individual or central body is able to fully understand the competitive forces of the market, and when it interferes it does so blindly, without a full understanding of the consequences. If a state intervenes in one area of the market it could have completely unintended consequences elsewhere. The role of the state is simply to create the conditions for the market to continue on its path of progress.

While holding him in high regard, Hayek's successors did not feel that he went far enough. *The Road to Serfdom* advocates against state intervention but it holds that some intervention is required in minimal cases, particularly in those cases where the decisions of individuals affect those around them. These include: various workers' rights like banning poisonous substances, limiting working hours, and ensuring minimal health and safety; the protection of the environment; the prevention of fraud; and the provision of some social and welfare assurances such as basic food, clothing, healthcare, and "a comprehensive system of social insurance in providing for those common hazards of life against which few can make adequate provision." He writes that "in no system that could be rationally defended would the state just do nothing."

For the neoliberal thinkers that followed, the state should do precisely that–nothing. This ideal was championed by Chicago School economists, including Frank Knight, the school's founder, and Milton Friedman, one of its most well-known economists. They took Hayek's ideas and pushed them even further. At the core of their teachings is the belief that market forces, whether they be supply or demand, unemployment or inflation, are akin to forces of nature. As forces of nature, they can be described using mathematical equations and predicted using calculations. This use of mathematics gives neoliberal economics a sheen of naturalism, presented with the same level of truth as the theory of gravity. It has allowed economists to present their ideas as matters of fact. Free-market calculations propose that if the market is truly free then it will achieve a state of perfect equilibrium. If there isn't a state of equilibrium, it's because there is too much interference in the markets. The Chicago School sought to achieve this utopian vision in which the market is truly free with all individuals acting in their own interests and no interference from the state.

To achieve a state of economic grace, Friedman proposed a three-pronged attack of deregulation, privatisation, and cutbacks. This meant privatising public assets which the state has invested in, selling them to corporations so that they can make a profit; removing any rules and regulations which affect the market, including price-fixing, minimum wage, and state legislation; and cutting back on any social spending on things like welfare, health, or other public services.

Today, neoliberal ideas have gone from being a theory proposed by a few sidelined economists to being the standard of common sense. To question them is to be out of step with reality. As with any dominant conceptual framework, neoliberalism has tapped into our instincts and intuitions in such a way that it feels innately *right*. Who would argue against a framework that insists on the centrality of individual liberty and dignity, safeguarding freedom, and the importance of self-actualisation? Starting in the 1970s, this became an attractive alternative to Keynesian economics and collective bodies and institutions who had lost their power and were finding it difficult to act. Neoliberal ideals took hold through a silent revolution that was imposed top-down by corporations and governments, under the auspices of individual freedom for all.

Despite giving itself the appearance of scientific reality, neoliberalism contains many inherent contradictions. For example, all neoliberal theorists are committed to minimum state intervention, except in those cases when the state defends the rights of private property, individual liberties, and entrepreneurial freedoms. However, state interventions are required to protect the market, as happened in 2008 when states had to bail out banks during the financial crises. As with many state interventions the main beneficiary of the bailout was the banks themselves. In the UK, for example, the government lost money when it sold off its share in the Royal Bank of Scotland.

Contradictions extend beyond state funding of private enterprise. Neoliberals have a great distrust of state power and espouse a philosophy of individual liberty, insisting that the state must stay out of the way of the market. This sits uneasily beside the coercive tactics and undemocratic methods that states will often take to impose neoliberalism and free markets; the prime example is Chile, where, in a CIA-backed coup, General Pinochet overturned the democratically elected Salvador Allende. The violent suppression of political parties, dissidents, and trade unions was all in the service of implementing free-market policies. To play on the words of Hayek, neoliberalism can be put into practice only by methods of which most neoliberals disapprove.

For the neoliberal, the role of the government is to create the right conditions for business to thrive. Across the world, neoliberal policies have transformed economies so that wealth rises to those who are the richest. Social inequality and declining conditions for the poorest in society are the costs of increased entrepreneurial risk and reward. As entrepreneurs and wealth-makers see their own lot improve, those who are the poorest or whose conditions worsen are blamed for their failure to increase their human capital. In the United States, since the widespread establishment of such policies in the 1970s, the share of national income of the top 1% of earners rose to 15%, with the top 0.1% seeing their share of national income rise from 2% to 6% between 1978 and 1999. At the same time the income disparity between CEOS and employees widened from 30:1 in 1970 to 500:1 in 2000. In the UK the wealthiest have increased their share of the national income from 6.5% to 13% since 1982. Worldwide, the income gap between the world's richest fifth of people and the poorest fifth of people has increased from 30:1 in 1960 to 60:1 in 1990 and 74:1 in 1997. In early 2017, Oxfam published a report that revealed that the eight richest people in the world have the same wealth as half the world's population. This group of eight men, which includes Microsoft founder Bill Gates, Facebook co-founder Mark Zuckerberg, Amazon founder Jeff Bezos, and Michael Bloomberg, between them hold half of the world's wealth. The exponential growth of this disparity between rich and poor has accelerated along with neoliberalism and free-market ideals, and looking at the figures it's hard to see neoliberalism as more than a framework for empowering social and economic elites, redistributing wealth from the poorest in society to the wealthiest. For some, this redistribution of wealth may be seen as a necessary evil to ensure some level of economic stability and security, but this point of view is only adopted by those – whether they be super-wealthy, moderately wealthy, or of middling incomes – who have benefited from such policies. The same cannot be said for those who have been negatively impacted, whether that's through the imposition of zero-hours contracts, the loss of jobs due to outsourcing, a diminishment in working rights and conditions, a tightening and reduction of the welfare state, the loss of social safety nets, or living through environmental disaster caused by the exploitation of resources. For those many whose quality of life has been greatly diminished through the unequal distribution of power and resources, there must be no comfort that there are some few who are living in luxury and extreme affluence.

Creating the right conditions for business to thrive does not just mean implementing policies which benefit businesses. It also means creating a society of individuals who accept those policies and see them as right, just, and necessary. It means creating a new type of subjectivity. For this reason, political figures who advanced neoliberalism were not concerned just with economic expansion. Margaret Thatcher, one of the key figures in the establishment of neoliberalism in the UK, was explicit about this in a 1981 *Sunday Times* interview. "Economics are the method," she said, "the object is to change the heart and soul." A goal of neoliberalism is to expand the market and the ethos of the market into every aspect of life. Society must be run along the basis of the market; everyone must be transformed into an entrepreneur, taking risks, seeking profit, acting individualistically, not just in their work but in every aspect of their life. To achieve this, the government created policies that nudged people into different ways of thinking. For example, in Britain in the 1980s, Thatcher introduced an enterprise allowance which individuals could opt for instead of unemployment benefits. The enterprise allowance provided them with

£40 per week during the first year of running their own business. The intention was to shift the mindset from being reliant on the state to being reliant on yourself, from a lifetime of state security to an individualistic enterprise culture.

Let's pause for a moment, take a step back, and reflect on what we're really talking about here: ideology and, in particular, the neoliberal ideology. The *Oxford English Dictionary* defines ideology as "a system of ideas and ideals, especially one which forms the basis of economic or political theory and policy." This is the most basic and neutral definition of ideology, and serves as a good starting point for any discussion of it. Wikipedia goes a step further, describing ideology as "a comprehensive set of normative beliefs, conscious and unconscious ideas, that an individual, group or society has." When we talk about ideology we are talking about the invisible structures which provide the normative framework through which we view society and culture. There are many different types of ideology, each with a competing way of defining the world, but there is always a dominant ideology and this dominant ideology provides the invisible justification for the way that power is structured.

Ideologies can be traced back throughout history, revealing themselves through the passing of time. In Plato's *Republic*, Socrates presents the idea of the noble lie–a lie which is used to maintain the structure of society. He talks of the myth of the metals: that when human beings were created they were each mixed with a metal–gold, silver, brass, and iron. The rulers are mixed with gold, their helpers mixed with silver, and the farmers and craftsmen mixed with brass or iron. The ruling elite are born to rule because they have gold in their souls, whereas the labourers and craftspeople must stay within the station defined by brass and iron. It isn't true, but it provides a justification for the tiers of social classes. It is a noble lie, or a useful fiction, which helps to maintain the social order.

History is filled with such useful fictions, foundational myths that give people an explanation for why the world is the way that it is. They are invisible to those who are living through them but become revealed to be fictions through the lens of history. For example, the fiction of the divine right of kings provided a religious justification for a monarch and the continuation of their dynasty. The belief that kings were chosen by heavenly beings or even imbued with some divine power seems archaic to us now, but for those living before the sixteenth century, when the concept started to be questioned, it was just an everyday fact. It gave legitimacy to the power of the monarch and maintained the Feudal system of which it was an expression. A more recent ideology which was given its full expression during the Industrial Revolution is separate spheres – the idea that women and men operate in distinct spheres of society. Men operate within the public sphere and women operate within the domicile. This division between the sexes was taken as simply the way of the world and, rather than being a form of oppression, it was seen to actually benefit women. Alexis de Tocqueville wrote in *Democracy in America* (1840): "although the women of the United States are confined within the narrow circle of domestic life, and their situation is in some respects one of extreme dependence, I have nowhere seen women occupying a loftier position." It wasn't until the mid-twentieth century that this ideology started to be adequately challenged and, while it remains the view of some individuals and groups, it is no longer naturalised as part of the dominant ideology.

These fictions, and others like them, are useful because they maintain the structures of power, usually in a way that benefits an elite. Because they are invisible, those who are controlled by them usually just take them as matters of fact: for centuries it was a fact that the monarch was bestowed his power by God, and for centuries it was a fact that men naturally occupied the public space and women naturally occupied the home. Through the lens of history we can see these for the fictions that they are. Many ideologies seem grotesque or anachronistic to us, but to the people living through them they were just part of everyday reality.

Through interrogating history, we can further refine our definition of ideology: an ideology is a fiction. While it is dominant, the fiction is taken for reality, but eventually all fictions outlive their usefulness and through the passing of time their untruth is revealed. It is difficult to intuit an ideology while you are living through it, but a general understanding of what an ideology is helps you to better question the conditions under which you are living. There are many different types of ideologies, or different fictions, all competing with one another, but there is always a dominant ideology. The purpose of the dominant ideology is to provide a rationale for why there is an uneven distribution of power and resources. We can explain to Plato's farmer that he remains a farmer because he has iron in his soul, to the Feudal peasant that he remains in his place because God wishes it to be so, to the Victorian woman that she belongs at home because it is the natural order of things.

Like the ideologies that have come before it, the objective of neoliberalism is to give a naturalistic explanation to the uneven distribution of power. It is different, however, in its form and its associated fictions. These include the freedom of every individual to act equally within the system; the imperative to be the architect of your own success or failure; the importance of business; the personhood attributed to corporations; small, non-interventionist government; the free market; the natural status of the free market; competition as central to all aspects of life; the centrality of choice; the inherent value of wealth; the link between social status and economic status; economic meritocracy; meritocracy in general; and the primacy of the individual over the collective. In time, through the passing of history, many of these fictions will be revealed to be modes of thought that just serve to obfuscate inequality and maintain the dominant hegemony.

In John Carpenter's satirical science-fiction film *They Live* we find a vivid metaphor for how ideology works, a connection that has been made by the theorist Slavoj Žižek. John Nada is a drifter, living on the fringes of a recognisable USA of the 1980s – the Reaganite, consumerist society that we are all familiar with. He finds a box of sunglasses and when he puts a pair of the sunglasses on the true state of the world is revealed: suddenly the ideology of the 1980s disappears and

Nada sees what's really going on. The world is ruled by an alien elite who keep the human race pacified through the use of subliminal messages in things like advertising billboards – obey, marry and reproduce, consume, no independent thought – and using an electronic transmitter which keeps people in a state of sleep. It covers over the substructures of society so that human beings accept the world as they see it and are unaware of their submission to the ruling elite.

When Nada puts on the sunglasses they scramble the signals that are sent out by the transmitter. The ideology is revealed for what it is – a fiction that maintains a specific configuration of power. We don't have to take the literal point from the movie that the earth is populated by a race of aliens who are keeping us as slaves – that would lead us into tinfoil hat, conspiracy-theory territory. Instead, the film is a metaphor that reveals the invisible illusion that the present is the natural order of things, that everything is just the way that it should be. It signals that there is a dominant ideology that invisibly structures reality for us. None of us will ever find an actual pair of glasses that reveals to us the underlying structures of the world; our glasses are the lens of history which discloses those fictions that have been crucial to maintaining the structure of power.

Why does this matter? Think back to that infographic by Founders and Funders – the one that speculates a future in which everyone is an entrepreneur. This is taken as a fact by so much of the tech industry and by society as a whole, but it is not a fact, it is an ideological statement and one that does not have absolute, necessary reality. It is a fiction, a useful fiction for some, but a damaging fiction for others, particularly those who have seen their working conditions diminish, their rights destroyed, their jobs sent offshore, and their security gone. It is not the case that in the future everyone will have to be an entrepreneur; this will only be the case if society continues with a dominant neoliberal ideology.

Through decades of state and media intervention, we have come to a point at which neoliberalism is not just an approach to thinking about markets, but is a framework for structuring every aspect of our lives. The entrepreneur, as its archetype, is a tool for governance. We must all be entrepreneurial. It is not just one path through life that we can choose, but one that we are forced to adopt in every aspect of our lives. As the 1980s progressed, and on into the 1990s, governments, economists, institutions, and corporations helped neoliberalism become the natural order of things: the individual replaced the collective, business replaced society. Through rhetoric that privileged freedom and individual liberty, neoliberalism restructured reality away from the post-war emphasis on equality and social democracy in a way that benefited the few.

Entrepreneurial qualities have become normative; they are prescriptions for living. They help you to be effective market agents. You must be more creative, more flexible, more innovative, more competitive. Around these ideas has grown a self-help industry which instructs us on how to be more entrepreneurial. Books like Philip Delves Broughton's *How to Think Like an Entrepreneur*, for example, which promises to "lead us to the heart of great entrepreneurial thinking; an understanding of our deepest human needs."

The dominant narrative establishes the entrepreneur and entrepreneurialism as objectively the best way to live your life. This creates a form of subjectivity that is able to cope with the precarious situation imposed by a market outside its control. Under a rubric of empowerment you can deal with the constant threat of unemployment because the responsibility for your unemployment is placed on you. It's okay to be unemployed because you live in a system where you can not only find a new job, but you are empowered to *create* a new job. You are responsible for everything that happens within your life–sure, bad things happen that are beyond your control, but it's how you deal with those upsets that matters. It's about your attitude: pick yourself up, identify new opportunities, move, grow, improve. Precarity is a good thing as it allows you to be flexible and answer market demands. Uncertainty is opportunity. When you become entrepreneurial you assert yourself as a subject within the market, you are a player, you are part of the game. Your success or your failure depend on your attitude.

The entrepreneur assumes uncertainty and risk and the entrepreneurial subject takes on this uncertainty and risk in every aspect of their life. Responsibility and risk are shifted from institutions, whether they be private or state, to the individual. Like the entrepreneur, you are expected to deal with high levels of uncertainty, finding opportunities for yourself in the ever-changing market. At all times you must be competitive, and if you fail it is no one's responsibility but your own. You weren't alert enough, didn't exercise good enough judgement, weren't flexible enough, fast enough. You must be the entrepreneur of your own life.

You should always make choices based on what will maximise your own interests, whether that's choosing hobbies, friends, career, or partner.

You are expected to strive for success in a system that cannot function without failure. You must be different. You must be flexible, bending and shaping yourself to forces outside of your control. You must accept working conditions that have no guarantees – zero-hours contracts, freelance and contract work, no security, no benefits; to ask for guarantees is an outdated way of thinking. You must always seek the new, shaking free from old ties and identities. You must continually grow and improve but never with any end in sight.

This entrepreneurial imperative neatly covers over the fact that any subject, any self, is not just an autonomous individual acting from their own free will, but is an assemblage of the non-individual–social practices, conventions, meanings, and institutions which we all share in common. Collective notions such as society, community, and solidarity are abandoned in favour of individualism, risk, and precarity, and justified by the heroic figure of the entrepreneur. Speaking of the economic and social changes that happened in Britain in the 1980s, the clinical psychologist David Smail writes:

> To move money quickly and easily, to dissolve obstacles in the way of rationalisation of working practices, and perhaps most essentially, to expand the scope and influence of the market, meant the wholesale and ruthless removal of as much as possible of the pre-existing social institutions and ideology identifiable as incompatible with these aims. This was to be the

Business Revolution, and in order to achieve its aims not only would existing business be enthusiastically recruited to the cause, but nonbusiness people would have to be - sometimes, but in fact surprisingly seldom, more reluctantly - re-shaped and retrained into becoming business people.

What happened was not just a change to business and economics, but a redefinition of reality along the lines of business and economics. This redefinition of reality continues in a stream of creative destructions and innovative disruptions as new technologies redefine reality again and again.

In 2011, a change in the format of the UK version of the reality-television programme *The Apprentice* underlined this shift in reality. Instead of competing for a job working with Amstrad founder Alan Sugar, candidates competed to be his business partner. The prize changed from a £100,000 salary to a £250,000 investment in the winning candidate's company. It is not good enough to be an outstanding, or even competent, employee. Let's not compete for secure jobs and stable salaries; now you compete for investment, you compete for an opportunity to roll the dice on yourself.

The only correct attitude is to be excited, to embrace it. The market is more than the flow of capital; it is an ethic. This is the culmination of the neoliberal project–everything is subject to the market. It has inveigled its way into every aspect of life so that the only way to participate in the world as an active subject is to be a consumer. A human being exists for the market, not vice versa. This ties your ability to act as a self-actualising subject to your income. If you don't have the money to participate you may as well not exist. Attempts to escape this ethic are branded by both culture and society as failures.

Instead, you must embrace the market and the market way of thinking, no matter that real value has been replaced by a marketing sheen that has all the substance of a perfume advertisement. Your ability to succeed is about your attitude, and the right attitude is one of hyper-excitement. Again, these processes were started in the 1980s. David Smail writes:

> Not only were the social issues and problems of every kind approached through the 'attitude change' mythology of advertising - everything from the training of the unemployed to the fight against AIDS - but even the most sober representatives of high culture found themselves declaiming the virtues of their wares in the manically urgent language of the supermarket. Works of art, scientific theories, affairs of state, medical treatments, courses of higher education, would all be sold with the same tired combinations of fatuous hyperbole. Everything was major, new (usually 'major-new' as a kind of compound attraction), unique, massive, important, 'important-new' and exciting. 'Stunning' and 'awesome' appeared a little later. Mediocrity was clothed as 'excellence'. Scarcely any social or vocational practice or pastime could be envisaged which did not seem both designed and expected to engender a kind of frantic excitement; the prescribed mode of mediocratic life was one of the mediation and consumption of euphoria, and anyone who attempted to engage in any other kind of activity,

or speak a milder or more considered language, stood in danger of finding him or herself beyond the boundaries of the real world.

This level of over-excitement and over-stating of reality has become reality. It has reached a zenith in the age of the internet. We are all hyper-connected and expected to be permanently excited about the launch of every "awesome" new "life-changing" product.

Today we are at the culmination of many years of state intervention which has resulted in the complete permeation of neoliberalism and neoliberal ideals into every part of society. We can see ideology everywhere we look, in the stories we tell ourselves, our popular culture, and our heroes and role models. Take, as an example, the film *Limitless*: Bradley Cooper plays Eddie Morra, a struggling writer who comes across a nootropic drug which allows him complete access to his brain power. He finishes his book, which his publisher loves, and then he gets on with improving the rest of his life. The possibilities, as the title implies, are limitless. Morra takes his new-found transhuman powers and, rather than going on to great literary and artistic heights, or using the powers of his unlocked mind for the betterment of society, he starts playing the stock market. The story it tells us is that when we have access to all of our brain, when we act in perfect accordance with our true nature, the best that we can do is increase our own personal wealth and power.

This narrative isn't just told to us in movies, but is a feature of almost every reality-television programme. In *The X Factor* it is not enough for contestants to be able to sing; they must also have a story, a place where they came from, an individual struggle that they have overcome in order to do what they were born to do (in this case sing). This sob-story narrative is prevalent on most reality television, perpetuating the fiction of individual struggle against all odds, with never any mention of the social factors that have helped the person succeed, let alone the fact that no matter what a person has struggled through in their past, it is their voice or their looks that will enable the recording company to sell their albums.

The centrality of business and entrepreneurship is propagated most obviously through TV series such as *Dragons' Den*, *The Apprentice*, and *Undercover Boss*. While *Dragons' Den* and *The Apprentice* promote the idea of the scrappy individual, *Undercover Boss* focuses instead on the benevolent boss. In *Undercover Boss*, an executive goes undercover in their own company to see how it is operating, make improvements, and reward individuals. Masquerading as an entry-level worker, the executive ends up in a variety of scrapes, is exposed to what is going on in their company, and learns about the lives of their employees. At the end, those employees are called to head office, there is a big reveal, and the employee is rewarded, usually amidst expressions of gratitude and floods of tears. The show tells us that executives are essentially good, rewarding hard work and ensuring good conditions for all. It contributes to the ideology of a business society as the best for all: any support should come from the benevolent private individual, not from the state.

Just as the media perpetuates the positive role models of ideology, it also perpetuates negative role models, people who we should definitely not aspire to be like, failures and abnormalities, people who are not entrepreneurial enough and therefore have failed. To fictionalise people on long-term benefits or who are unemployed would be seen as mocking the poor; instead they are presented under the auspices of reality television, in the whole and true presentation of themselves. A familiar example is *The Jeremy Kyle Show*, which sees the feckless, the drug-addled, the alcoholic, the poor, the work-shirkers, the unemployed, brought onstage to be berated and humiliated by Kyle. Often these people suffer alcoholism and drug addiction, or mental-health conditions such as schizophrenia or depression. No mention is made of the wider social conditions or context which led to their appearance on *Jeremy Kyle*; the narrative is just about them and their degeneracy. People who watch *Jeremy Kyle* run the ideological gamut of those who look down on his guests as scum, to those who watch with horror, to those who roar with laughter, in many cases accepting what's on the screen as a straightforward presentation of fact.

The other dominant "poverty porn" format is the reality-TV show that gives viewers a look inside the lives of the poor and those claiming benefits. These shows are as numerous as the more aspirational reality-TV shows that seem to dominate television schedules. In the UK, Channel 5 and Channel 4 lead the way with shows like *How to Get a Council House*, *Benefits Street*, *Benefits Britain: Life on the Dole*, *On Benefits and Proud*, *Benefits and Bypasses: Billion Pound Patients*, *Benefits by the Sea: Jaywick*, and *The Big Benefits Handout*. These shows, often voiceovered by working-class or northern actors, made by production companies staffed by the upper-middle classes, present people on benefits as something other, something to be looked at for our entertainment, amusement, and horror. They uphold the dominant ideology that claiming benefits is aberrant and those people who do so are a scourge: the only thing left to do with them is have them on television, for our entertainment. Like Kyle's show, they tend to avoid the social, economic, and political conditions that have led to a person claiming benefits.

All poor narratives have their holes, and while reinforcing the dominant ideology, all of these shows have their cracks and fissures that show the points at which ideology breaks down, revealing itself as a fiction. *The Apprentice* shows us how the upstart can strive and succeed and take control of their own life. But no one watches *The Apprentice* for that. The joy of watching *The Apprentice* comes from the sheer idiocy of the contestants, whose bombastic claims are a million miles away from the reality of their talents. We enjoy watching these entrepreneurs be foolish, argue, screw up: it never really matters who wins because they are invariably mediocre. A show like *The Apprentice* does not show how difficult starting a business is, does not fulfil the narrative of the individual doing it for themselves, but reveals how straightforward business actually is and the types of overblown, under-talented people who often pursue it. Equally, shows that seek to vilify the working classes and those on benefits often end up revealing their subjects to be thoughtful, caring, and intelligent people, who display camaraderie and community spirit, but who are trapped by circumstances beyond their control. These ruptures and slippages show just how thin the narrative of neoliberalism really is.

While the ideology of neoliberalism has many supporting characters, its protagonist is the entrepreneur. It is the entrepreneur who acts in perfect harmony with the system, is the engine of change, helping it to achieve its goals while profiting (or failing dismally) along the way. The fiction of the entrepreneur is a fiction that many live their lives by: the imperative to take risks, to be independent, individual, different, to break the rules, to speculate, to change, to be flexible and opportunistic. This is not just an imperative for those in business, but for everyone, at all levels of society.

It's not hard to pinpoint the place at which the narrative of entrepreneurship breaks down. You can see it both in those who are encouraged to always be entrepreneurial, and in actual, dictionary-definition entrepreneurs. The breaking point of entrepreneurialism, one of the costs of the dominant ideology, is depression. With all risk and responsibility placed on the individual it's no surprise that depression is the great mental-health epidemic of our time. The WHO estimates that 350 million people worldwide suffer from depression. And this number is growing: in 2012 the BBC reported that the number of people in England living with depression had increased by half a million in three years. While not uniquely produced by capitalism, mental-health issues associated with distress are much higher in more developed countries, particularly those with a dominant neoliberal ideology; countries in which public services are privatised, there is little regulation of financial services or labour markets, which are individualistic, and which associate freedom and choice with the freedom to choose within the market. In 2004 a worldwide WHO mental-health survey found a wide disparity in the incidence of mental-health illnesses across various countries, ranging from 4.7% in Nigeria and 4.3% in China (Shanghai), to 20.7% and 26.4% in the United States. It's not a huge leap to look at the socio-economic factors that might cause such a large incidence of mental-health illnesses in the US.

The psychologist Oliver James has emphasised how the increase in rates of mental distress correlates to the rise of neoliberalism. He cites studies done by Joseph Veroff and others in 1957, 1976, and 1996. Each study was carried out in the United States and as they all followed the same methodology, all are useful for looking at the increase in distress over time. The question "Have you ever felt that you were going to have a nervous breakdown?" was put to respondents. In 1957, 17% said they had; in 1976, 19.6% said they had; and in 1996, 24.3% said that they had. The biggest increase was between 1976 and 1996 and broadly ties in with the widespread introduction of neoliberal policies in the United States.

In the UK large representative samples of people in their thirties and early forties who were born in the same weeks in 1946, 1958, and 1970 were asked about their emotional wellbeing. Again, there has been a big increase in depressive symptoms that ties in with the introduction of neoliberalism: of those women aged thirty-six in 1982, 16% reported "trouble with nerves, feeling low, depressed or sad"; in 2000, 29% of thirty-year-olds reported the same. A further study showed that between 1977 and 1985 reports of psychiatric morbidity (e.g. panic attacks, phobias, and depression) increased from 22% to 31%. Taken in aggregate, these studies begin to show a correlation between changes in society and economic outlook, and individuals' mental health.

There is a body of psychological and social research that investigates the impact of setting goals and aspirations based on extrinsic factors, such as financial success or status, or intrinsic factors, such as self-actualisation. In a 1993 research paper entitled "The Dark Side of the American Dream," Tim Kasser and Richard Ryan studied the difference in psychological wellbeing between students who pursue the extrinsic goal of financial success vs those who pursue the intrinsic goals of self-actualisation and community. They found that students who pursued financial success were struck by "less self-actualisation, less vitality, more depression, and more anxiety." There have been subsequent studies into Kasser and Ryan's claims, but they broadly follow the same results: that having extrinsic motivations to pursue extrinsic goals leads to less fulfilment and happiness, and greater psychological distress.

This correlation between extrinsic motivations and distress is perhaps why we've seen such an increase in the number of people who are depressed or suffering from mental illness, particularly in English-speaking Western countries such as the United States and the UK. While there is evidence to show that pursuing money for the sake of intrinsic factors (such as challenging oneself or cultivating a stable home life) can bring happiness, studies show that pursuing money and status as an end in itself does not. And yet we are brought up in a culture with the imperative to acquire wealth and status–this is an imperative that we find not just through the media and celebrity, but through government rhetoric, schooling, and internet culture. This is sadly apparent in massive funding cuts both to the arts and theoretical sciences, and in the downplaying of such subjects in our education system. If it's not vocational it's not worth doing. And yet, any extrinsic goal that we aspire to is contingent upon factors beyond our control; therefore we lose control of our own lives and the ability to shape our own destiny. Despite advocating liberty and the freedom of the individual, all of the ideological imperatives of neoliberalism are based on extrinsic factors: be more flexible and bend to the market, self-actualise through being a market actor, be more alert so you can react to market changes. This way of living your life, buffeted around on the tides of the free market, cannot lead to happiness.

There are, undoubtedly, entrepreneurs who decide to become entrepreneurs for intrinsic reasons. These could be for the pleasure of intellectual enquiry, for the challenge of growing a business or taking a product to market, to build a community, or to do some social good. But it is the nature of the market to disrupt this process. As soon as a startup takes venture capital it is beholden to its investors. Investors do not put money into a business just so that the entrepreneur can achieve her goals. They do it to increase their wealth. This means that, whatever the motivations of the entrepreneur themselves, the startup's goal is to make money. Eventually, the venture capitalist needs to see a return in their investment in the form of an exit, either when the company is sold to another company at a profit, or through an IPO (Initial Public Offering) when a company is successful and profitable enough to sell stock to the public. This means that in the startup world, no matter what your good intentions or intrinsic motivations, you will always end up subject to the market, pursuing extrinsic goals for the benefit of others.

I am reminded once again of Adam Smith's "invisible hand" theory, that when individuals act for themselves they benefit society more than if they actively seek to benefit society. Perhaps if we lived in a society where everyone was free to pursue what intrinsically mattered to them, whether that be knowledge, or family, or community, self-actualisation, or peace, or identity, this might be the case. But our reality comes from the dominant ideology that originates outside of us, that is beyond our control, and that, like Hayek's free market, we can never really fully comprehend. This ideology is invisible, has been given the appearance of the natural order of things. We take it for common sense and act in the ways that common sense dictates. And yet our knowledge of it is always inadequate; without understanding those forces that act upon us we are unable to make rational choices. When we pursue extrinsic goals we are passively driven by market forces, and so despite what we are told about our individual liberty, we cannot be said to truly be free. The theory of the invisible hand only works in a society without ideology, but that society cannot exist.

While the prevailing culture puts pressure on every individual to take more risks, be more flexible, go out on your own, work every hour, and pursue your business single-mindedly, the people who take this to heart and live it as a creed are entrepreneurs. This has resulted in a macho culture in which you must be seen to be working all of the time and if you aren't working you are partying. While it might be unfair to characterise all in this way, the stereotypical "bro" culture that has come to be synonymous with startups hasn't come out of nowhere. Many startups apply a "work hard, play hard" ethos, with rampant discrimination against minorities, especially women, no rights for employees, and poor working conditions such as no breaks, long hours, and no time off. And even in those who don't fall into this category, the culture of every startup is to work as hard as possible, to compete, and to raise capital.

It's not surprising, then, to see that the global trends in depression are reflected by the number of actual entrepreneurs who are sharing their experience of mental-health issues, particularly depression. In recent years, more and more successful startup founders and venture capitalists have written about their experience of depression. There is a growing number of entrepreneurs who are speaking out about depression and mental illness. In 2011 Ben Huh, the founder of the Cheezburger network, wrote about his struggles with depression. In the article, he talks about the failure of his first startup in 2001. He writes, "loneliness, darkness, hopelessness... those words don't capture the feeling of the profound self-doubt that sets in after a failure. Loneliness. Darkness. Hopelessness. Those words describe the environment of depression. Self-doubt? That shakes you to the core and starts a fracture in your identity that makes you question if you should even exist anymore."

Looking around him, Huh saw other entrepreneurs who were enjoying success, raising investment, and going public. This solidified the feelings of self-hatred that he was plagued with. It was nine years later that he secured a $30 million investment for his company from investor Brad Feld. The Cheezburger network was a success story, and Huh's experience shows how an entrepreneur can rise from the mire of depression, but for every success story there are countless others who fail.

Around 90% of startups fail within three years. Some of these founders will go on to found other startups but many will not. As a coping mechanism, the startup culture has embraced failure as part of its mythos. The adage "fail fast, fail often" is bandied around like a mantra, and many entrepreneurs wear their failure as a badge of honour. There is even a conference around failure–FailCon–whose mission is to help "founders learn from and prepare for failure, so that they can iterate and grow fast." But while failure might be part of the reality of running a startup, no one really wants to fail, and the valorising of failure seems to just be a way to deal with this unpalatable side of entrepreneurship. When failure actually strikes, the costs, both financial and personal, can be so high that the founder sinks into depression. There have been a number of prominent cases of startup founders taking their own lives.

Startup founders who fail experience distress, but so do those are successful. Achieving financial success will not necessarily bring happiness. As discussed earlier, a person's happiness is affected by whether they pursue intrinsic or extrinsic motivations. There are many startup founders who set out with intrinsic motivations. However, society's definition of success is usually determined by extrinsic factors and the tension between the two could be cause for distress. Achieving the measure of success as defined by others does not necessarily lead to happiness. In 1999, a review of studies of intrinsic rewards vs extrinsic rewards found that in 128 studies, when extrinsic rewards increased, intrinsic motivation decreased. It may be that the pot of gold that sits at the end of the entrepreneurial journey leaves a person feeling empty.

Indeed, many of those startup founders and entrepreneurs who have been most vocal about depression are those who fit the criteria for success. In an article for the *Guardian*, Niall Harbison describes how the solution for him was to "take medication and work through therapy to get over it and manage it." This approach may treat the symptoms and provide the individual with coping mechanisms for future attacks, but it doesn't treat the causes – the wider social and cultural landscape that has led to such a huge amount of pressure, self-doubt, and loathing in the first place.

Much of the research on whether entrepreneurs have better mental health and wellbeing is contradictory: some say that they suffer from greater mental-health issues, others that they have better mental wellbeing. These discrepancies might be explained by the fact that we live in a society in which everyone is expected to be entrepreneurial, and it's impossible to remove actual entrepreneurs from this context. It's easy to blame the things that are right in front of you for your failures. The most obvious culprit is yourself: you failed because you were inadequate, you were unable to do what was required, be more flexible, take more risks, succeed. In order to do better next time you must work on improving yourself, you must pay more attention, be more alert, move faster. This attitude ignores the fact that we exist in a system that conditions us to feel this way, and that there are ever-changing market forces beyond our control.

Entrepreneurialism has risk and uncertainty built into it at its core. You cannot be an entrepreneur without a high chance of failure. This is compounded by the necessity to always be alert, to exercise good judgement, and to be ahead of everyone else. When things start to go wrong, even when it is caused by factors outside of your control, it's hard not to blame yourself. If market changes mean that a product you've invested heavily in doesn't succeed, then it's your fault for not being more aware of those changes or not being quick enough to act. Despite a slim chance of success, you take on a huge amount of risk, for yourself and your mental wellbeing, for the people you employ, and for the investors whose money you are playing with.

Success is interpreted in terms of your entrepreneurial virtues, failure is attributed to your personal failings. Anything systemic is excluded. Your success is not because of your social status, nor because of your class, because of the contacts and network of your parents, because of public-school confidence. You did not succeed because you got a small business grant, or because you took advantage of state-funded research. You did not succeed because of who you know but because of who you are, because of your perkiness, your spunkiness, your independence, your alertness, your entrepreneurial spirit. You did not succeed because you just happened to be in the right place at the right time. You did not fail because you went to an overcrowded and underperforming school, or because your parents didn't have the education to school you at home. You did not fail because shifts in the market meant you lost your job and you've been unable to get the skills to get a new one. You did not fail because a competitor broke into a market and had more financial backing than you. You did not fail because illness has meant you have lagged behind. You failed because you weren't fast enough, weren't adaptable, flexible, alert. You failed because you weren't entrepreneurial enough, it's because of you, and no one else.

A sign that the transition to an entrepreneurial culture is complete is the increase in the number of people starting their own businesses, particularly young people. According to social entrepreneurship foundation UnLtd, 55% of young people want to start their own business. In 2006 145,104 under-thirty-fives founded companies, up to 247,049 in 2013. Many of them have a "killer idea," something that they think will disrupt the market. However, most startups fail and the biggest reason for this is not due to some personal failing but because there is no market need for what the startup is selling. It's not enough to have a great idea, you have to have an awareness of the market. We have created a culture in which we encourage young people to think that they just need one idea to succeed, in which anyone can do anything, when the reality is that more often than not their idea leads to failure.

The fiction of the entrepreneur has become for many people, especially young people, the only common-sense way of measuring success. It's one-sided, therefore, that school and society rarely paint the entire picture. The story goes that success is due to one individual. But what's not talked about are all of the other factors that contributed to that person's success. One of the archetypes of the modern entrepreneur is Steve Jobs, the co-founder of Apple, who is lauded as the success story and hero of modern times. Many of the articles about Jobs that litter the internet focus on the various personality traits that he had, and that you should emulate, that made him a successful entrepreneur.

There's no denying the major role that Apple and Jobs have played in changing both the technological and social landscape, but the image that is portrayed of Jobs – that he was a trailblazer, out doing it himself, that Apple did it on their own – is incorrect. There is an element of individual genius about Jobs – his creativity and dogmatic commitment to simplicity and his ability to think laterally were part of the company's success. However, much of that success was based on technological innovations that came about through state research and funding. As Mariana Mazzucato covers in detail in her book *The Entrepreneurial State: Debunking Public vs. Private Sector Myths*, Apple has been successful because of the support of the state, not in spite of it. The company received early finance from the US government's Small Business Investment Company programme. Also, it spends a fraction on research and development in comparison to companies such as Microsoft and Google. Instead, Apple has been highly successful at combining and exploiting technologies that emerged through state investment. These include the internet, GPS, their touchscreen display, Siri, battery technology, and GMR. And Apple isn't the only hugely successful company that is reaping the benefits of public investment: the research for the algorithm behind Google was funded by the National Science Foundation, for example, and Tesla founder Elon Musk's companies, including Tesla Motors, SolarCity Corps, and Space X, have received $4.9 billion in government subsidies. These companies are happy to take full advantage of state investment in technology while the state gets no share of the rewards (many of these companies go out of their way to avoid taxation). Technological progress does not happen in a bubble; it accumulates through time. A new technology does not arrive in isolation but is built upon all that preceded it, and many of these technologies are created through state-funded research.

Mazzucato's research gives lie to the myth of the entrepreneur as purely a self-made person, an independent agent, a successful, lone individual. The underlying reality and the fiction it supports are inconsistent. You are told to survive on your own, to be self-sufficient, achieve highly without the support of the state, yet those who are the highest achievers, our modern-day heroes, are assisted by state subsidies, tax cuts, and government funding, reaping the rewards of government investment. No one can ever be the archetypal entrepreneurial hero because that person does not exist. It is a mythical creature produced by the dominant ideology.

The dominant neoliberal ideology has found one of its most effective expressions through the technological revolution of the internet. As politics and economics have operated on us, so has the internet, and both have changed how we live, work, and relate to one another. The internet is a playground for entrepreneurs and the entrepreneurial attitude. During the early days of the World Wide Web the internet was rightly seen as a space for the counter-cultural, a place for bottom-up self-organisation, where communities and collectives could form. However, it is the nature of neoliberalism to marketise everything, and the internet–with its free flow of information, its emphasis on freedom and emancipation–was an easy medium to subject to the market. Like neoliberalism, it advocates ideals of freedom, individualism, openness, and emancipation. It is built on interconnectedness and on speed, on change and on disruption. It creates individuals within a network, moulding them through nonhuman forces that implicate everyone within the system.

Just like neoliberalism, the internet operates with a set of contradictions. This was not inevitable but is a result of a technology emerging at the same time as a specific ideology. The ideal of neoliberalism is that all individuals have the same access to the market in which they can self-actualise; the ideal of the internet is that all individuals have the same access to information. But, just as neoliberalism excludes many people from the process of self-actualisation, on the internet access to information is disparate.

The first barrier is that people have different technologies with which they access the internet. Some have high-speed fibreoptic broadband, others operate on mobile networks. In Q1 2016, Akamai reported that the average global internet connection speed is 6.3 Mbps: at 26.3 Mbps, South Korea has the fastest average internet, followed by Hong Kong at 20.1 Mbps, whereas Pakistan has an average connection speed of 2.0 Mbps, and Yemen 0.7 Mbps. The speed of a connection determines how fast a page or resource loads – as internet speeds have become faster, so web designers have been able to increase the size of web pages. People in South Korea will find their pages load instantly, while in Yemen they will have a very long wait. Of course, this assumes that people can access the internet at all. Many individuals won't be able to access the information superhighway since they are unable to afford an internet-enabled device or internet connection.

In August 2013, Facebook tried to address this situation with the launch of Internet.org, and later with their "Free Basics" application. This followed on the heels of a white paper published by Mark Zuckerberg asking the question: "Is connectivity a human right?" He argues that the internet gives access to what he calls the global knowledge economy. The laudable goal of Internet.org is "bringing internet access and the benefits of connectivity to the portion of the world that doesn't have them." The company wants to provide basic internet access to anyone who has a smartphone, thus increasing internet access to people in developing countries. In partnership with a huge number of telecom operators they have launched the Free Basics application in fifty-three countries in the developing world. In order to make this financially sustainable, Facebook restricts access, only giving access to websites that use small amounts of data – this means websites that are text-heavy like

Wikipedia and dictionary.com, the Bing.com search engine, and Facebook itself.

While the intentions appear commendable, the practical application goes against the principles of net neutrality: that Internet Service Providers should provide equal access to information, treating all data the same, and not discriminating or charging differently because of factors such as operating system, user, platform, content, type of equipment used, or mode of communication. Basically, that all people have equal access to information. When people connect to the internet using an application like Free Basics they only have access to websites chosen by Facebook. They have access to the Bing search engine, for example, but not to Google. They allows videos, but not from YouTube. Critics have said that it is part of Facebook's plan to confound people into thinking that the internet and Facebook are the same thing. This isn't as far-fetched as it sounds: in focus groups in Indonesia, a researcher found that while individuals said they don't use the internet, they do use Facebook. For them the internet doesn't exist, just Facebook. With the widespread adoption of Free Basics, Facebook is in the position of being able to define what the internet is for the poorest people in society. This means excluding competitors and creating a narrow funnel through which the World Wide Web is viewed. As a technological innovation, Free Basics perfectly reflects the realities of neoliberalism: it's permitted to restrict some freedoms in order to protect and promote the rights of business. And all this is done in the name of promoting freedom and individual liberty.

It isn't just companies that curb the freedoms

promised by the internet. As with the markets, where it suits their private interests, corporations are constantly putting pressure on governments to increase regulation on the internet, decreasing openness and freedom. There are ongoing attempts to prevent individuals from sharing files such as films, music, and games. And governments and other bodies legislate in order to curb individual freedoms online. The UK's Digital Economy Bill increases sentencing powers for copyright infringement and gives the government powers to fine mobile network operators and pornographic sites that show certain acts. This violates the principle of openness on the internet, which is supposed to provide free access to all information. In the UK a visit to a torrent or streaming site will just lead to a standard message saying that the site has been blocked by a high court order. Those who are technologically competent enough to use a proxy or a browser like Tor are able to access those websites, but it excludes the majority of the population.

It's on the internet that we see the pragmatics of the tensions within neoliberalism played out. On the one hand, openness, freedom, and minimal state intervention; on the other, a constant lobbying for states to intervene where business interests are in danger. It's no surprise that many internet purists are libertarians or techno-libertarians, espousing the pure openness and freedom of the internet. In this world, copyright is an outmoded concept and patents stifle true innovation. Attempts by corporations to curb freedoms of openness and interconnectedness through government lobbying show that they pay lip service to the free market and reduced state intervention only when it suits them. When it doesn't, when they need laws in order to maintain their power and control, they can throw their money and influence at the government to bend market flows to their own needs.

I am convinced that the potential of the internet cannot be fully realised when it is subject only to neoliberal ideals and market forces. That path leads to further regulation in the interests of business, and an internet that is primarily a tool for corporations and individuals to make money. We will arrive at a place where the spontaneity and creativity of the network is eradicated by restrictions that channel flows only towards supporting and bolstering the market. This eradicates any space for collectivity, creativity, and emancipatory politics.

Like the free market itself, the internet as playground of technology and ideas is only open to the few, not the many. As we promote a global philosophy of individual freedom and responsibility–which insists that any inability to succeed is a personal failure, which celebrates the wealthy and pays large financial bonuses to the very people who destabilise markets–around three billion people are living on less than $2.50 per day. These people have no failing other than being born into conditions which make it impossible for them to participate. And while worldwide aid efforts are made to reduce poverty, they are done within an international system that ignores corrupt business practices that rob these countries of a trillion dollars.

Recall the warning issued by Hayek in *The Road to Serfdom*. Hayek argues that, because we lack a full awareness of the market, central planning and socialism always have unintended consequences that lead to totalitarianism and serfdom. We are at a point in the neoliberal adventure when we can see the emergence of its own unintended consequences. Even in my own lifetime, the transition to hyper-individualisation has wrought changes that could not have been predicted. The insistence on individual choice and liberty has created a vast disparity between the rich and the poor, and widespread deregulation has had, and continues to have, catastrophic economic and environmental consequences.

But what of us Western millennials? Those who have taken to heart the experiment started by our parents' generation; those who have inherited the bounties of the free market and whose individual freedom surpasses the freedom held at any point of history?

In early 2017, the freelancer marketplace Fiverr ran a new advertising campaign. Fiverr is part of the gig economy, where freelancers can sell their skills starting at $5 per job. In 2017 the company launched an advertising campaign called "In Doers We Trust," which captures something of the reality for people who work in the gig economy. One of the adverts, which was widely shared on social media, shows a bedraggled-looking millennial with the words: "You eat a coffee for lunch. You follow through on your follow through. Sleep deprivation is your drug of choice. You might be a doer." A doer is a go-getter, someone who will do anything to succeed, someone who forgoes their own health to make some money and to get ahead. The reality of the gig economy, and its associated freedom, is that many gigs are so poorly paid that those who participate do have to work every hour of the day to make enough money to get by on. Companies like

Fiverr celebrate the fact that the people who make it viable, who are working all of the gigs, do so to the detriment of their own health, with no security and no benefits. A press release about the campaign, cited in the *New Yorker*, says "The campaign positions Fiverr to seize today's emerging zeitgeist of entrepreneurial flexibility, rapid experimentation, and doing more with less. It pushes against bureaucratic overthinking, analysis-paralysis, and excessive whiteboarding."

This narrative is perpetuated everywhere, but particularly in the gig economy. There is a constant stream of success stories that are supposed to inspire other workers, to make you think that pushing yourself to the limit for cash is aspirational. Another much-cited story came from ride-sharing company Lyft, who published an effusive article on their blog, about Mary who was driving while nine months pregnant. She went into labour during a shift but she stayed in driver mode and answered a ride request while en route to the hospital. She continued to work, despite the fact that she was having contractions. It's sad for Mary that she had to keep working while she was in labour, but even more unsettling is that Lyft, a company that provides no healthcare or maternity benefits, thought that it reflected well on them.

On the one hand, companies like this capitalise on their workers' success stories, on the other they refuse to give them any of the basic rights of employees. Food-delivery company Deliveroo has created a form of doublespeak in order to get around the fact that their workers act like employees but they don't want to treat them as such. A six-page document describes how managers should refer to workers: instead of a "hiring centre" they have a "supply centre," instead of "shifts" they talk about "availability," instead of "uniform" they have "kit" and "equipment," instead of "clocking on" they "log in," instead of "earnings" they have "fees." These linguistic gymnastics are just a way for the company to avoid paying even the most basic statutory benefits such as holiday, sick pay, or national insurance.

This attitude extends beyond the gig economy and into other tech startups. Canadian-based startup SkipTheDishes cancelled a woman's interview because she sent an email asking about her pay and benefits. In the email cancelling the interview, the company said "Your questions reveal that your priorities are not in sync with those of SkipTheDishes," i.e. the company expected the woman to care more about working for them than any remuneration she might get for her time and work. In a follow-up email they said, "We believe in hard work and perseverance in pursuit of company goals as opposed to focusing on compensation. Our corporate culture may be unique in this way, but it is paramount that staff display intrinsic motivation and are proven self-starters." SkipTheDishes is not unique in this way. Anecdotally, I have heard of a number of CEOs who will not continue with a candidate's application if they ask about compensation too early in the process. Many startups expect dedication to the company to take precedence over concerns such as feeding yourself or paying the bills.

We live in a culture that says these are matters of choice: if you don't want to work for Deliveroo, don't. If you don't want to be in labour while you are driving a taxi, start your (unpaid) maternity leave earlier. If you don't want to work every hour of the day you don't have to. However, research by the Pew Institute in the USA shows that for 60% of people who use labour platforms such as Uber and Deliveroo, this type of work is either essential or constitutes an important part of their income. The proliferation of gig companies and the sharp rise in the number of people who are either participating casually or for whom it is their whole income, indicates that the path that we are on is not one in which people who carry out labour are treated fairly.

The road that we are on may not be to Hayek's serfdom, but what lies at its end is not, except for a few, wealth, affluence, and freedom. This road is one on which everyone walks alone, one of overwork and underpay, one in which individuals are nominally free but are bound by necessity to a company that refuses to grant them even the most basic rights. It is a road littered with casualties, and far from the equilibrium and bounty promised by the forefathers of neoliberalism. If we pause on our march and take some time to reflect on where we are going, it's not too late to mitigate some of the worst repercussions.

In 2017, after thirty-eight years of neoliberal policies, it's possible that we're beginning to catch a glimpse of what's at that road's end. Neoliberalism's failure was marked by the 2008 financial crash and followed by a long period of austerity politics, during which we were spun a narrative of belt-tightening, a fiction that led to cuts to public spending and the devaluation of our incomes while corporations enjoyed as much, or greater, profit as ever. In the UK, blame for our troubles was misattributed to either the Labour Party or the impoverished–welfare claimants, refugees, and immigrants–rather than the banking and financial classes. People just aren't buying it any more. The narrative that concessions must be made to business to protect jobs and the economy is flimsy indeed when there is mass unemployment, people struggling on zero-hours contracts, and the economy itself has failed. People are using their votes to reject neoliberalism and we are coming to a fork in the road.

In 2016 it felt as though the crushingly inevitable conclusion to neoliberalism was a swing to the right. In rejecting the establishment, many voters in the UK and the US have moved towards right-wing populism, as signified by the election of Donald Trump in the US and the vote for Brexit in the UK. This is echoed across the world as more and more countries and nationalities are becoming insular, protectionist, and xenophobic. People want someone to blame for their problems and right-wing elites have used the opportunity to blame "liberal" elites, along with minorities such as the impoverished and refugees, for their problems.

But it doesn't have to be this way. A divided society–in which the rich get richer and everyone else is distracted by squabbling–is not the only way. Neoliberalism has successfully created a generation of entrepreneurial subjects, who, unlike the generations before them, are highly risk-tolerant, highly networked, and highly adaptable. Its natives are the generations of millennials and post-millennials who have been brought up immersed in this ideology and who are extremely technologically savvy, able to see and unlock technological potential. Those of us from these generations are not only equipped to be effective market operators but we get technology and know how to use it. We bear the brunt of neoliberalism and austerity politics, and, over the past ten years, have been waking up to the disparities and unfairness created by the neoliberal agenda. As our parents enjoy their retirements, we live in a world of insecure jobs, uncertainty, and no promise of stability at the end of our working lives. Is it possible that neoliberalism has created the perfect agent for its own destruction? One that is dexterous, technologically-minded, and alert. Is it possible that the next creative destruction will be to the free market itself?

It's not historically unprecedented that an ideology produces the very form of subjectivity that ends up overturning it. Those pre-Enlightenment scientists such as Copernicus and Galileo who, under the patronage of the church, pursued scientific endeavour for the further glorification of God, developed theories and conducted research that led to the scientific revolution. Their discoveries meant the eventual overturning of the dominant ideology of the church and the divine right of kings. Has the dominant neoliberal ideology self-generated a subjectivity with the capacity to undo it from within?

The signs are there. Already in 2008 the Obama campaign effectively used social media and a message of hope to oust the Republican Party in the US. But as things have become more tumultuous, the ideological narrative thinner, more frayed, the disparities in society even greater, technology is being used to overturn accepted norms. Despite a pernicious mainstream media that branded Jeremy Corbyn a terrorist sympathiser, a clown, and unelectable, during the 2017 UK election campaign there was a greater swing to the left than in any election since 1945. Technology and social media were used as the tools of collectivity, demonstrating the power given to collective action by technology. It seemed unthinkable only a year ago that the electorate could vote counter to the mainstream media; it is not only possible but it is happening.

It's true that internet usage has grown along with neoliberalism and in many ways it has been co-opted, but the two are not intrinsically linked. As the narrative of neoliberalism fails, the liberatory potential of the internet remains, available to a subject who has innate understanding of how to live and survive in free-market capitalism but who rejects it wholeheartedly. Maybe it's only at this point that the true emancipatory potential of the internet will be unleashed. This makes it more important than ever to resist those attempts made by those in power, whether they be governments or corporations, to place restrictions on the internet that make it solely another tool for the market rather than a tool for forms of collective action and community organisation.

I have spent a lot of time thinking about that infographic from Founders and Funders. It sums up a lot of what I have come to realise about entrepreneurialism: that there is a sheen that both contributes to and hides all of the inequalities at the heart of our society. By promoting a world in which everyone is an entrepreneur, the infographic promotes a world of great inequality, where it's acceptable for just eight people to hold more than half of the world's wealth, where it's better to have no safety nets, where everyone fends for themselves, where people skip lunch and become sleep-deprived to work, where they continue to work through labour contractions, where they can fall through the cracks and be forgotten.

There is a troubling contradiction at the heart of entrepreneurialism on the internet: we are told that we are individuals and free agents, and yet companies and designers are always seeking to change our behaviour. We are told we are individuals by the people who want to mould us, who want us to act in certain ways. I am suspicious of an ideology that one the one hand tells us that we are individuals, but on the other uses techniques of behavioural psychology to change who we are.

In the end, I decided not to pursue running my own business, instead getting a job at a company. I don't want to be an entrepreneur, nor do I think that in the future everyone will have to be an entrepreneur. That is a viewpoint that needs to be resisted – entrepreneurialism should never be all that there is. It is just one way of being among others.

Thrilled as I initially was by the fact of earning money from my writing, and despite revelling in the joy of having escaped having a boss, I was finally turned off by the thought of growing it as a business. Like many other people I took what I loved and turned it into a business. I did what so many self-help books and gurus tell us to do: I took my passion and I made it into my job. But the reality of this is that you take what you love and you subject it to the market. I remember the excitement I felt when I started to make money from writing – I thought "hey, people actually value this.

People will pay me to do something I care about." But even though I cared about writing, I didn't actually care about what I was writing about. I was writing for the market rather than writing for myself. A decision to build and expand upon my own business would mean pushing that even further, probably to the point where I wasn't doing any writing at all, just managing other people.

I am fine being an employee. I am fine with not being an entrepreneur. I don't need to launch my own business in order to feel like a success. There is more to life than acquiring VC funding, launching a product, disrupting a market, or making an exit. Just because the dominant ideology says that you have to take risks and innovate doesn't mean that you do. It's okay to just earn enough and do the things that make you happy, preserving what you care about from the market and keeping it for yourself. It's okay to give as much to your job as it requires, reserving some of your energy for your family or your hobbies. It's okay to do things that make no money but have their own value that no one in the world understands except you

PART THREE

NAVIGATING THE INFORMATION SUPERHIGHWAY

CHAPTER SIX
PANOPTIC PEOPLE

Just when I was considering whether I should expand my writing business, a job offer came my way. I had been working online for about three years. During that time I had got to know many companies that operate fully remotely. It's not just freelancers who are able to take advantage of working online, it creates a different work landscape for employees too. I didn't need to fly solo: I could enjoy many of the freedoms of working online without the pain. I recall the decision-making process when I went to my first remote job. I decided that growing my own business wasn't for me. I wasn't interested in dedicating all of my time to establishing and growing a company–my priorities were elsewhere. My options were to continue freelancing or join a company. Joining a company would mean a stable, reliable income every month, no more admin work, no more invoicing, no more looking for clients, less task-switching, more focus, and a company to pay my expenses. The trade-off for this security would be that I had less freedom and I would, once again, have a boss. What swung it was that I no longer wanted all of my time to be monetised. I wanted my work time to be my work time and my non-work time to be free. With a stable income every month I would no longer be thinking about the money I could be making with my downtime. This, I hoped, would give me more time to devote to my relationship and the other things that matter in my life.

I went from the scattered and often manic world of remote freelancing, to being a member of a remote team. Unlike independent freelancers, remote teams are highly interdependent on one another, often working together on the same project and relying on one another for information and work product. There are many different ways in which companies operate remotely: some maintain a large head office while allowing employees to telecommute, others have offices scattered across the world, and then there are companies that are fully remote, having no office, or only a small one, with its workforce fully distributed. Since giving up on my freelance writing business I have worked for two fully remote companies, first as a contractor, then as an employee.

All of the remote companies I have worked for have given me a high level of freedom, allowing me to take on and manage my own workload, work when I want, where I want, and be task-driven rather than working to a clock. There may be an element of luck at play, but it's also true that this level of freedom is necessary for a remote company to operate properly. It is impossible to monitor and track remote workers using traditional methods, so many companies focus on hiring people with the right drive and skills to work independently.

For this to work, a successful remote worker must demonstrate many of the qualities associated with entrepreneurialism. As a remote worker you must be assertive, creative, collaborative, and take responsibility for yourself. Many remote workers have little direct oversight from a manager; this means that you have to take the initiative and be able to work independently. You need to be self-driven, and able to be productive without any direct external stimulus. You must know when to ask the right questions and who to ask. You speak up with ideas and don't allow yourself to become forgotten. You are able to collaborate with just the online communication tools that are available. You are creative in finding the right tools and right processes to get the job done. You take on a high level of responsibility for yourself. Remote companies work because remote workers have internalised the skills they need to make those companies work.

Remote companies shift much responsibility from themselves onto the individual. Whether a company works hard to maintain good conditions for its workers or not, it is inescapable that the balance of risks and costs is weighted more heavily against the employee. There are the tangible costs such as paying for an office, including utilities like heating, internet, and electricity. Then there are the intangible costs—relying on employees to prioritise their own tasks, to work independently, for pastoral care. With their heads down in their work, working long hours, often much longer than a co-located employee, remote workers assume an additional responsibility for ensuring that they have proper, face-to-face contact with people. Some remote companies do little to address this, others explore ways that they can shift risk and other burdens back from the worker to the company.

Being remote offers many benefits to companies, including having lower operating costs and more productive employees. One of the biggest positives for a company is that it can hire people from anywhere in the world. This benefits both the company and individuals working for the company. Companies get more access to talent, and individuals have more access to job opportunities. Hiring can be done on the basis of skills, not on location, and individuals don't have to turn their lives upside-down in order to relocate for a job. This opens up job opportunities for people all over the world and in any location, whether that's in a major city or in the middle of the countryside. It increases work options for people who are disabled or experience things such as environmental sensitivities, mobility impairment, episodic symptoms, chronic pain or fatigue. It also increases opportunities for work amongst people with caring responsibilities, giving them access to work or careers they couldn't have otherwise. For employees within a company, it provides a space for greater cultural enrichment as people from different countries and backgrounds get an opportunity to interact.

For employees, there are many benefits to working remotely and it's true that working for a remote company provides a lot more freedom than working in an office, and that if one's time is properly managed it's possible to create a good work/life balance. However, remote employees face many of the same challenges as remote freelancers: isolation, overwork, difficulties creating boundaries. Added to that are the challenges that come from working as part of a remote team, including difficulties with communication, time zones, and guilt.

Remote companies need to hire people who thrive on independence, which means that they need to create a culture based on freedom. In order to do this, they often focus on results instead of time spent working, and allow employees to work when they want and where they want. A growing trend in tech companies is the unlimited holiday policy. Employees can take as much time as they want, provided they deliver work and perform. This is a huge perk. Who doesn't want unlimited holiday? Unlimited paid time off looks great on paper. It creates a policy that is applied equally to everyone in the company, people can take time off if they are getting burnt out or are sick, and it is an opportunity for employees themselves to strike the right balance between work and life.

An open-holiday policy only works for a company, though, if it employs people who are motivated, self-driven, and who enjoy their work. They are often people who are so dedicated to what they are doing that they find it very hard to take time off. Some people don't take any time off at all. When I started work at one company I was talking to a colleague who had been in his role for eighteen months. I was enthusiastically talking about the open-holiday policy, already making plans for what I would do and where I would go. I asked him how much time he had taken off since working there. His answer: none.

One of the reasons for this echoes what I found as a freelancer. If you are at home and you enjoy your job, you might as well be working. The lack of physical boundaries between work and home make it very easy for the two to become blurred. As a result, people work many more hours than they would in a traditional office job. People with the highest level of responsibility or who are predisposed to overwork are particularly disadvantaged by unlimited holiday policies because they find it hardest to create a gap in their workload.

The effectiveness of an unlimited holiday policy is dependent on the culture of the company. The company may talk about freedom, but the standard for how much holiday is taken is set by the company leadership and the employees themselves. If no one is seen to be taking time off then it is hard for anyone to fully take advantage of the policy. If everyone is working 24/7, no employee wants to be seen to be the one who isn't pulling their weight.

It can cause resentment amongst employees if some people are taking more holiday than others. This is complicated further in globally distributed teams where team members have different attitudes towards time off. In Europe, for example, it is normal for people to have four to six weeks holiday per year, but in the United States, where holiday is usually two weeks per year, this is seen to be excessive. What an unlimited holiday policy effectively does is make the employees themselves responsible for setting the standard of how much time is acceptable to take off. Some companies don't see this as a problem – if employees don't properly take advantage of their unlimited holiday it is their own responsibility. Other companies mitigate the negative impacts of such policies by having a minimum-holiday policy; in others managers will step in to ensure that people take time off.

An increasing number of companies adopt a "flat" approach to management. They are non-hierarchical, or as non-hierarchical as possible, without traditional management structures. A traditional pyramidal organisation has a head at the top and power functions down through the organisation. Many companies are getting rid of this structure in favour of flat organisational structures which "empower" employees, allowing them to be self-driven and self-determined. The medium of the internet allows companies like this to function particularly well – individuals, responsible for themselves and their own work, are distributed, often across the globe, answerable to themselves and their peers. Power no longer flows up and down but is distributed across the network. This new approach to business structure seems like a triumph of democracy, a flattening out in which everyone can be the boss. I'm not sure, though, that it adequately solves the problems associated with middle-management and bureaucracy. When everyone is empowered to make decisions, it's not always clear who should be making the final one. This can cause confusion amongst employees, who sometimes just need a straight answer. Also, there's a danger that decisions will be overturned and railroaded by a CEO or other higher-up who isn't happy with what's happening and who's the only person with the authority to step in.

As with much of the operation of a remote company, this approach shifts responsibility from the company and onto the worker. Each and every worker is responsible not just for themselves and their work product, but for the smooth running and success of the company. Remote companies only work because the ideal remote worker does not need a high level of external surveillance and monitoring. This doesn't mean we require none, but that we have internalised the techniques of discipline required in order to make us into effective workers.

While writing this book I've reflected on what motivates me and ensures that I get work done. I've gone through periods of manic obsession with productivity, but even since I've moved beyond that I constantly am looking for ways to work better and do more. I worry both about being productive and appearing to be productive. Where does this internal drive towards self-discipline come from? What drives it? Who is this person that I need to be productive before? I find myself returning again and again to the image of Jeremy Bentham's Panopticon as written about in Michel Foucault's *Discipline and Punish: The Birth of the Prison*.

The Panopticon is the design for a prison. The building is circular, with cells around its circumference. In the centre is a tall tower where a guard monitors the inmates. Because of the circular design of the building, the inmates in the cells are always visible to the person in the central tower, and yet the inmates cannot see who is in the tower. They do not even know when someone is there watching them. Even in the absence of a guard in the central tower, surveillance is permanent: if you do not know when you are being watched, you must always act as though you are being watched. The design of the building trains the inmates to act in a specific way, to behave.

In *Discipline and Punish*, Foucault uses the Panopticon

to describe how power works. For Foucault, power is not coercive, it is not just a way for me to get you to do something. Power is the way that different bodies, whether they be individuals, institutions, collectivities, or whatever, act upon one another. In the analogy of the Panopticon, power comes from both the structure of the building and from the presence of the different individuals within the building. The interaction between these factors determines how the guard and inmates act. This is how society functions: we are caught up in a complex network of power relations that determine how we act and that shape us, creating our identities and our subjectivity. Actual surveillance isn't necessary if an individual has already internalised the systems of surveillance.

I've spent so much time agonising over productivity techniques, berating myself about my inability to work to some nebulous standard – a standard that I don't know the origin of. No one stands over me to tell me to work harder and work faster, no one but myself. And yet I can look back at my own upbringing and see the training and structures that have led me here. From the moment I started school I was subject to ringing bells that told me when to act and timetables that told me where to be, homework diaries that kept track of everything I had done, a uniform that prescribed how I looked. I was tested so that I could be ranked alongside my peers. I was taught a standardised curriculum that inferred that knowledge is objective and unquestionable. My teachers, parents, and other grown-ups were the eyes of authority. When I started to work I was subject to the timesheet, to shifts, and the rota, and eventually I slid, like so many people, into the nine-to-five. These are all tactics of power, the bricks and mortar of the Panopticon, that helped to shape me into a worker ready for the world. Just like the inmates in Bentham's prison, I don't need to see the authorities in order to act as though they are there.

All of these specific tactics of discipline are part of the wider power networks of technology and ideology. I may have internalised the systems of surveillance needed to work independently, but my subjectivity has also been shaped to operate in the fast-paced, fast-changing world that I have found myself living in. I look at the pace of life for previous generations, one of jobs-for-life and the promise of stable pensions, and perceive it, rightly or wrongly, as a slow, steady, march, while my world feels punctuated by catastrophes and changes that I am expected not just to weather but to thrive on.

This make it easy for companies to give people like me so much freedom. I come preinstalled with a set of rules that guide my action and allow me to discipline myself. I have the tools to get work done and I can cope with high levels of change, risk, and uncertainty. I may not be an entrepreneur but I know how to act like one.

Recently, I had a conversation with a colleague about our company's minimum-holiday policy–we have a twenty-eight-day minimum per year, and can take as many days in addition to that as we want. Although there is no limit, it feels like there is a limit. Once I have taken twenty-five days I start thinking "can I really take any more holiday? Don't I need to save some days for Christmas?" Or, as my colleague said, "I just feel so guilty about it."

We don't need to feel guilty; the company encourages

employees to take as much holiday as they want. So where does this feeling of guilt come from? It doesn't come from others in the company, it doesn't come from leadership. It can only be internal to ourselves, the residue of years of training to act in specific ways.

Feelings of guilt are not uncommon amongst remote workers, and they aren't just related to holiday policies. I experience guilt regularly and try to fight against it, knowing that it only comes from inside me. But there are factors that contribute towards its appearance. The first comes from the fact that no matter where I am, I always have the tools at my disposal in order to get work done. It doesn't matter how much work I have done in a week, if someone else is reliant upon me or I have a deadline looming, there is a constant pressure during my periods of downtime. I feel responsible for getting work done so that other people aren't held up, or I feel like I must push myself as hard as possible to meet a deadline.

And then I find that I judge myself against those people around me. Remote workers are judged by their work product. The most obvious way of looking like you are busy is through communicating. Some workers are constantly communicating, constantly sharing, constantly busy and productive and doing great things. It's hard not to judge yourself against these people and feel that you are always coming up short. Because we do work online and rarely see one another, it's impossible to know what the reasons are for these people's productivity. Maybe they're working eighty hours a week, or they have no other responsibilities they have to worry about. Maybe they're on the verge of burnout. Maybe they're not doing a lot of work at all but they just appear to be busy. We judge ourselves against others without knowing any of the conditions that they work within. There is a huge amount of anxiety that comes from just not being able to see people that you work with regularly. There's always a space for second-guessing, for assuming what people will think, for projecting.

Working with teams across a wide variation in time zones can make feelings of guilt and responsibility even stronger. Time zones are a big challenge for globally distributed teams. My usual time zone is GMT/BST. My team mates, however, work everywhere. My daily cycle looks something like this: wake up in the morning and say hi to the Australians–they are winding down for the day and some are getting ready for bed. Chat throughout the day to people in Europe, by lunchtime people in the US are starting to appear, first of all those on the east coast and then, by the time my working day ends, people on the west coast. If I need to work with them or talk to them about anything I may extend my working day long into the evening in order to work with them in real-time. In order to be in touch with all the people I need to speak to, I have to be online or available for most of the time I am awake.

I have found that it can be hard to create and maintain relationships with team members in vastly different time zones. It's not impossible, but it's certainly more difficult than maintaining day-to-day chatter with people who only have a few hours' difference. Most friendships develop when people are online together, and while you can read through scroll-back it is impossible to directly participate. For some people this means extending their working day in order to get that connection. In big teams or companies it can result in geographically located silos that are grouped around time zones. This can be compounded by the realities of the work that a company is doing. If a company is client-focused, it may be necessary to form a team around the time zone the client is in. And there are the practicalities of working together across time zones. When a team member needs information from someone else who is in a different time zone, they may have to wait until the end of the working day for that person to wake up.

It struck me a few years ago on a trip to Portland, Oregon, that not everyone finds it necessary to extend their working day. I sat down in the evening, opened my laptop, and there was no one else online. My virtual world had fallen silent, and evening, which was usually such a busy time for me back in the UK, was suddenly quiet. I powered down my laptop, watched Netflix, ate, and fell asleep, thinking about how the online world we experience is determined by the physical locale where we live. Our online worlds are created and defined through the connections that are available to us during the time when we are actually online.

Modes of communication have become increasingly fast-paced so it is harder to communicate asynchronously. Ten years ago, when blogs were prolific, being online at a specific time didn't matter quite so much. The mode of communication created by blogs is slower than lots of today's applications. Someone crafts a post, and commenters take time to respond. Things could happen in a flurry but not of necessity. Today, much communication happens at higher speeds–people work in real-time chatrooms, Twitter-storms erupt out of nowhere, and being online and being present at a particular moment matters. Therefore time zones matter.

To mitigate the challenges of remote work, companies have meetups or other ways for employees to come together. Meetups provide an opportunity for all-important face time between employees. A meetup, loosely defined, is a short period of time during which a company, or a team within a company, gets together in person. The purpose of the meetup may be to work on a project, but often they just serve a social function, providing an opportunity to get to know other members of the team.

Meeting up in person can have a huge effect on how you subsequently relate to someone online. There are often broad language and cultural differences between people working at the same company. A sense of humour, irony, or sarcasm may not come through in text; the person might just seem blunt or rude. I can think of many people who I have disliked online, but when I meet them in person they have a sense of humour and a set of mannerisms which simply don't translate well to the online medium. After meeting in person, it's always easier to communicate online, easier to crack jokes and have relaxed banter. Video calls are more natural, they're like returning to the space when you were all together. When you read someone's text you can hear their voice in your head, imagining the rhythms, cadences, and inflections of what they are saying.

Depending on the company that you work for, a

meetup could be anywhere. Nationally distributed companies meet up in the same country, internationally distributed companies meet up anywhere in the world. There are drawbacks to working in this way though. Remote workers often spend large amounts of their time at home by themselves. A meetup requires that you travel and spend up to a week in intense contact with a group of other people. For people who are more used to isolation, this experience can be both invigorating and draining. I go away from a company meetup with an enriched sense of my team and colleagues but always leave exhausted and need time to recuperate.

Meetup culture privileges certain groups of people, usually abled-bodied, healthy people, those people in countries that don't have difficulties getting a visa, and those without caring responsibilities. If you are caring for someone, whether it's a child, a partner, or a parent, it can be difficult to get time away for a meetup–either you don't want to be away from your family or you are unable to. This can undermine one of the huge positives to remote work–that it gives such people access to work that wouldn't traditionally have been available to them. If a company requires three or four meetups a year this puts a huge extra burden on the person. While working from home allows caregivers to work and provide care, it can be difficult to participate fully in a distributed company. This has a disproportionate impact on women: a 2014 US study found that women are twice as likely as men to look after elderly parents; and women, whether they are a primary carer or they work and look after the kids, spend more time on average on childcare than men. In contrast to men, women are also more likely to cite childcare as a major motivation for choosing to work remotely. This can make it more difficult for women than men to get away for a meetup. This disparity is another expression of the pre-existing structures of power which we find ourselves living in and which are often unreflected upon. The remote company appears to liberate women, touted as a way for women to continue their careers while looking after their children, but within that is an unspoken acknowledgement that women should do double the work, both rearing children and working, now at the same time.

It's impossible to work online and not see the way that women and other minority groups experience prejudice. From online harassment to unconscious bias, being a woman in the tech industry involves struggle. While I have experienced the usual thoughtless comments about my looks and my skills, the most pronounced way that I have experienced the gender imbalance is through the physicality of motherhood and how it impedes women and not men. In 2012 I was at an event in Arizona, chatting to a man whose wife had recently had a baby. He was at the conference, drinking until the early hours of the morning and having a lot of fun. He missed his new son but there was no physical barrier to him being away from home. Every time I met a man at a tech conference whose wife was at home with a baby it struck me again. I realised what a big impact the gender difference has during those early years of a baby's life.

Fast-forward four years later and my own baby was four months old. I was organising a conference in

London. It wasn't far from home but I needed to be away for three nights. I was breastfeeding; in order to be away from him I needed to express milk. I calculated that he would need around one litre of milk per day, so around three litres in total. I knew when he was born that I would be taking this trip, so I started to hoard breastmilk in the freezer. I could express around two hundred millilitres per session, which took about thirty minutes. To pump three litres of milk I spent, in total, around seven and a half hours pumping, to take a three-day trip. This was unpaid labour time, carried out during maternity leave, in order to be away for work. Beyond the time was the invasiveness of pumping, of feeling like a cow being milked.

Then there was being away itself: when a baby is feeding a lot in the early days it needs to feed about once every few hours. Otherwise your breasts fill up and you can get engorged, which can lead to mastitis. This meant I had my pump with me for the trip, and every few hours I had to disappear to use it. I also had to wake up in the middle of the night to express. I would go to my hotel to pump, wake up in the night to pump, go home from the pub to pump, find a quiet room somewhere to pump. While I appreciated that I had been able to escape my baby for a few days, it wasn't without its problems. I was pushing it longer and longer with my breasts turning into massive painful rocks until I'd sorted it out.

I wasn't aware that my pump, though not cheap, was more suited to once-a-day pumping than pumping seven times in a day. Over the duration of the three nights the amount I was able to express became less and less. My breasts weren't emptying fully. It became painful to pump. To express milk you need to stimulate a let-down – this is when your milk literally comes down – and in order to this you need to be relaxed. I was not relaxed. The less milk came out, the harder and more painful my breasts got, the more tense I got, and the circle continued. The last night I could pump hardly anything. At 1am, while everyone else was out drinking, I was sat in a hot bath, desperately massaging my rock-solid breasts in the hope that something would come out. I spent hours with my breast pump, upset and exacerbated, up until 4am for just a few millilitres.

Exhausted and emotional, I noticed that one of my nipples was looking strange. With each pull of the breast pump, painful lumps on my areola were filling with fluid, swelling up and forming angry blisters. The more I pumped, the bigger they got. Angry red blebs, pumping up and down, filling with fluid and blood. I switched of the pump and cried myself into an uneasy sleep, hoping that in the morning I wouldn't have mastitis.

When I woke, I phoned D and told him not to feed our son. I needed to do it the instant I got home. I rushed for the train and was back in the space of a few hours, my breasts in agony. The instant I put my baby to my breast the milk came down and I felt a wave of relief. But when I put him on my left breast and he latched on I screamed. There was a horrendous eruption of pain as his latch broke the blisters on my nipple. And it happened every time he latched on over the following days.

I persisted with breastfeeding, not wanting a work trip to mean that I had to stop. I just put up with the pain. I used ice to freeze my nipple before a feed, I took endless painkillers, used a wooden spoon to bite down on as he latched on. I was in a double-bind: I had to feed him or I would get mastitis, but every time he latched on it was making things worse. I was terrified of my pump, as it had caused the problem. Eventually I went to a breastfeeding councillor who suggested I rent a hospital grade pump. I was still nervous about using it but I used it on my left breast until it had healed.

All that, just for three days away from my baby. Next time, I would bring him with me or not go. So when I meet men at conferences with very young babies, I'm reminded of the very real, very physical, and very painful reasons why women end up excluded from things like conferences and meetups. The WHO recommend that women breastfeed for the first two years of a baby's life, which makes travelling very difficult. Meetup culture, like so much of society and the tech industry, is a priori set up with the needs of the status quo, which is usually white men of a certain social class. These are the power structures that we are living within and with which society is woven. They are not explicit but are the underlying assumptions that we make about the world. Therefore when we make decisions about things that seem trivial, whether that's how many meetups to hold a year, travel expectations for a particular job role, or the hours that we expect a person to work, we often do them by applying blanket assumptions that only actually work for one group of people. It may not be intentional but, like so many imbalances, it requires conscious intervention.

As it becomes more common for companies to operate

remotely or at least offer telework, we are learning more about the changing landscape of work and how it impacts employees. Some companies already see the negatives as outweighing the positives. Big internet companies like Google and Yahoo discourage or ban telecommuting amongst employees. This isn't purely because they are stuck in the dark ages, as some proponents of remote work would have you believe. Research carried out using wearable sensor technology found a relationship between face-to-face communication and creativity, and a further study found that programmers who are co-located produce fewer code dependencies (dependencies are considered bad in software engineering). There are also varying degrees of knowledge-sharing between co-located and virtual teams: co-located teams have the highest level of knowledge-sharing, followed by virtual teams who have met in person, and followed by virtual teams who have never met.

Work provides a source of connection and answers a need to belong. Many people draw their identity from their work and this is harder to do if you are part of a virtual company, particularly if that company makes no effort to create stable work-role identities. There are, undoubtedly, those individuals who work from home because they prefer to work from home – they prefer the isolation and they don't want to work closely alongside other people. However, proponents of buzzwords like "the future of work" see all work moving in this direction, whether someone is suited to it or not. Virtual workers miss out on many of the things that people value about work: water-cooler conversations, lunch breaks together, drinks after work. They also lack many of the markers of a company that help them to form an identity; things like dress, or the layout or design of an office. For people who are not suited to it and whose employers do not provide that stability of identity, individuals need to look to families and local communities to strengthen other modes of social connection.

The biggest hurdle, or series of hurdles, for any remote worker or remote company, is communication. The first hurdle is that, when a job goes from the office to online, the workload of the person doing that job increases. In addition to their work, every online worker has an extra communication burden: they have to communicate nearly everything they do in order to be able to effectively do their job, almost to the point of over-communication. Communicating online is not just the act of writing. The individual also has to make the effort to ensure that they are understood, navigating all sorts of hidden cultural and linguistic differences across the team. Despite this additional burden, which can take up a considerable part of a person's day, there is rarely any decrease in that person's expected output or work product. This contributes to the fact that people who work remotely or telecommute are more likely to work long hours, to the benefit of employers and the detriment of employees. The increase in workload is just one aspect of the challenges of communication, signalling a minefield of potential problems for the remote worker.

CHAPTER SEVEN

WEARING GYGES' RING

I was on a trip with a client, let's call him Jim, who I had just started working for a few months previously. I had joined a small team and I was excited that this was the first chance I'd had for a long time to work closely with other people. I had completed a substantial piece of work about a month earlier, which I had sent to Jim, awaiting his feedback. It was a big piece of writing, heavily researched, with a lot of thought and time put into it. It was also the first project I had completed in my role and I was still unsure of myself. I was nervous, I was new to the job, wanted to impress, wanted to feel that all of the hard work I had done had paid off. Since we hadn't yet spoken about it, I assumed that we would sit down and talk it through when we were together in person.

When we arrived, I asked Jim if he had looked at my work. He hadn't, but he said that he would. This made me nervous–I wanted to be able to relax but couldn't until it was out of the way. A few days into the trip and the team was sat around a big square dining-room table, our faces turned down, bathed in the blue glow of our screens, the room silent except for the staccato tap-tap-tap of fingers on keyboards. Through the kitchen door, framed in the open doorway, Jim sat at a high-top table, reviewing my work. I was anxious that he was reading it, but glad that we'd finally be done.

Time passed with everyone absorbed in what they were doing. A small red notification symbol appeared on the Skype logo on my toolbar. I clicked on it. Messages were appearing in our team Skype channel. They were messages from Jim – his feedback about my work. I glanced over to him, confused. He was reporting his thoughts, his back to me, his face turned to the screen. I wasn't quite sure what to do. Should I respond in the channel? Should I go and talk to him? I felt immediately panicked, confused by this detached way of communicating.

I looked up. Some of my colleagues were looking at me. One of them rolled their eyes, as though familiar with the scenario, another looked back to his screen and a new notification symbol appeared on mine.

"Well, this is awkward," he wrote.

I looked up at him and raised my eyebrows.

We all watched as the feedback came in, terse and non-specific, without empathy or nuance. Jim was talking to me but the room was in total silence. He acted as though I was completely elsewhere. I felt a rise in tension, confused, unsure what to do. It wasn't the feedback itself but the manner in which it was given that upset me: impersonally, in a way that ignored that I was right there, in the flesh. In those moments there were layers and layers of communication, online and offline, data passed over the internet, glances with unspoken understanding, a silent rise in tension.

When he was done he stood up, closed his laptop, and went upstairs, to shower or take a nap or something. I wanted to follow him and say "I'm right here," but I didn't, because he was the client, I had just joined the team, and I felt I had to play by his rules. He didn't talk to me about it afterward, it was as though he had had that conversation with someone else, as though my online self was a different person to my physical self, that in person we could be pals but online he could act without feeling.

The incident made me reflect a lot about how we communicate with one another online and offline, and how we navigate the space in between. It became, I suppose, symbolic for me in that it captures what it means to privilege text-driven communication. Text-driven communication is used to keep others at a distance, to reject empathy, to create a gap between your online persona and your offline self.

I am as guilty as anyone else of choosing to send a text message over picking up the phone. Since that experience, I try to deal with any more difficult issues over voice chat, video, or in person, no matter how stomach-churning they make me feel. Text has become the default mode of communication for many who work online, but nothing can replace looking at one another, face-to-face.

On the internet we communicate just with text. We embellish our text with punctuation, with emoticons and emoji, but we can't capture all of the nuance of face-to-face conversation. When you talk to someone in person you are presented with their face; you look them in the eye and you are accountable for what you say and how you say it. Another person's face elicits a challenge, a demand that you act with humanity, empathy, and compassion. Delivering difficult news, being cruel, participating in a confrontation – all these things are made more difficult when we have to look someone in the eye. But that doesn't mean that we should avoid doing them. Part of being a social human being is engaging with those things that are difficult as well as those things that are easy. If we shift all of our difficult communication into a zone that lacks empathy we contribute to a world that treats people in only the most utilitarian fashion. Treating another person as a person is important both for their humanity and for your own.

In the transaction of text-based communication what matters is not the person, but the information that they send to you. Perhaps this is why I felt so affronted by Jim's approach to a difficult conversation – the conversation undermined me, placed the information ahead of my own personhood. And yet, in any communication, there are at least two participants, and those participants are human beings who are more than the words that they type into a keyboard. So many times I've found that online interactions lose an element of humanity. Communication is reduced down to bare facts.

Text-based communication may appear easier, may appear to be more low-friction, but it introduces many challenges that you don't have if you are sitting in a room talking to another person. Firstly, whoever you are talking to is distracted, more so than someone communicating on a medium like the telephone. When you converse with text, both people are on some level distracted: writing emails or code, responding to their other messages, looking over social media, reading the news, making a playlist, writing a book. I know from my own work online that I'm rarely just having one conversation. I usually have a few direct messages going on as well as open-ended discussions in group chatrooms. If a conversation is important enough to need all of my attention I try to jump on a call–voice or video. But even then I find my gaze looking over Twitter or scanning chatrooms.

When we converse with one another online, we come laden with the implicit rules of offline, face-to-face conversation. They're built into us. If someone were to stop speaking in the middle of a conversation, we would, from the context, be able to tell if they were angry, thoughtful, upset, happy, overcome. Online, text-based conversations are filled with pauses and, since we're unable to gauge what someone is feeling, we fill those pauses with our own anxieties and projections. There is no way to gauge the impact of what you say other than to wait to hear what the other person says. You can't see them smile or frown or wince.

Every text-based conversation is punctuated by gaps. I send a message to someone and it might be a few minutes or even hours or days before I get a response. Despite knowing that the person I'm talking to is as distracted as I am, this gap makes me anxious, particularly in important conversations. I say something and then I wait and I wait and in my mind I'm running through all of the different things that they might be thinking. Usually it's a worst-case scenario: I am not getting a response because the person is thinking about how to tell me something bad, about what an asshole I am, or how pissed off they are with me. Some tools, like Slack and Skype, have a feature that displays a message which says "so-and-so is typing." This means I know that I am getting a response, but it can be equally anxiety-inducing and I run through all sorts of scenarios in my head: why are they taking so long to write? They must be outlining, in detail, how wrong I am and how stupid. Why did they write for ages and then post nothing at all? They must have realised that what they wrote was so devastating that they couldn't post it. Why have they written for ages and then only posted a few words? They can't have told me everything that they are actually thinking. That long, drawn-out pause often means that the other person hates me and is going to tell me how stupid I am, that I need to reassess my life and stop being an asshole. And, by the way, you're fired. Kthxbai.

Text-based communication has lots of scope for misinterpretation. Even with native English speakers, it can be difficult to understand any phrasing and cultural idioms that are not shared. Through communication with bare text, as much as we try to convey ourselves, meaning is lost, and there's no surefire way to know if it has.

Words take on different meanings across cultures, idioms lose their sense. As a native of Northern Ireland, I am steeped in a culture rich in local idioms, many of which make no sense to anyone outside of Northern Ireland. Words like "scundered," "culchie," "boke," "geg," and "hoke," and phrases like "shut yer bake" or "yer ma's yer da" make sense to speakers of Ulster English, but not much to people outside the country. Over the years, since I've left Northern Ireland, I've found myself flattening out my use of dialectic, partially to make myself more understood, but also because I've found it embarrassing to sound so provincial.

This is not a new phenomenon: in Thomas Hardy's novel *The Mayor of Casterbridge*, the novel's eponymous mayor, Michael Henchard, chastises his daughter, Elizabeth-Jane, for (what the author calls) "her occasional pretty and picturesque use of dialect words – those terrible marks of the beast to the truly genteel."

> The sharp reprimand was not lost upon her, and in time it came to pass that for "fay" she said "succeed"; that she no longer spoke of "dumbledores" but of "humble bees"; no longer said of young men and women that they "walked together," but that they were "engaged"; that she grew to talk of "greggles" as "wild hyacinths"; that when she had not slept she did not quaintly tell the servants next morning that she had been "hag-rid," but that she had "suffered from indigestion."

Elizabeth-Jane's use of dialect disappears due to her father's embarrassment that she speaks like the farm labourers and workers, rather than the daughter of a Mayor. I have equally been embarrassed about the way that I speak, so over time my vocabulary has become more homogenous, more like everyone else, and I realise when I get back to Belfast how different my speech now is.

When I write on the internet I have to make a

conscious effort to dampen it down even further. Not only do I have to get rid of those Northern Irish turns of phrases but I have to even be cautious using idioms that are in common use in the UK. Equally, I am confused when people use idioms that I am not familiar with. It's even more complicated when speaking to people who don't have English as their first language. Then you have to be sure that you are communicating what you mean in bare language, as clearly and as accurately as possible. When we use cultural idioms that aren't shared by everyone in a group, a burden is placed on those who don't understand. They have to look up the meaning, often leaving them second-guessing and feeling stupid.

There are numerous studies that show how people from different cultures communicate in different ways: in a study of American and Greek students, the American students thought that the Greek students were too social and the Greeks felt that the Americans were too self-oriented; a study of work performance in computer-mediated communication (CMC) amongst Turkish and US students found that hierarchical status had a greater effect on Turkish students; and studies of instant-messaging conversations sent by Chinese and American participants found argumentative structure to be culturally specific. There are so many hidden but meaningful aspects to the way that we communicate. These can affect all of our online relationships and create space for misunderstanding. If I don't have an implicit understanding of the way another culture operates it's inevitable that all of the hidden meaning conveyed by a person will be lost. Communication in text has a level of danger and requires a level of caution that is much easier to overcome in direct, face-to-face communication.

Through the necessity of being understood, we are forced to drop our "occasional pretty and picturesque use of dialect words." Language loses its poetry, its richness, and its chiaroscuro. We have to select those words that are most clear and precise in all contexts and to all people, and lose those words that are the signatures of our cultural difference. We communicate in a way that is flat, losing the aspects of language that make us unique. For the sake of comprehension we give up all of our linguistic idioms for the utilitarian purpose of communicating information.

Beyond our local idiosyncrasies, so much that is joyful about conversation does not carry to the online medium. We lose the rhythm and flow of a direct, face-to-face conversation; half-sentences, cutting in on top of one another, finishing off each other's sentences, pauses, losing a train of thought while speaking, staring off into the distance, the shared search for the perfect descriptive word. We lose the sound of another person's laughter at a joke, their exasperated sigh, the raising tone of their voice if they are angry or excited, the way that they fidget or their eyes dart from side to side. We cannot tell if what we've said has made them smile or frown, made them feel wistful or longing, cannot tell by the cast of their face whether they are ready to crumple into tears or erupt with joy. No internet tool, whatever its tricks, can replace the richness of a conversation in person.

In the online medium, it is difficult for things to be otherwise. Ambiguity in written conversation can lead to confusion at best, and conflict at worst. The people who struggle the most online are those who find themselves unable to say what they want clearly and concisely. Words are presented in their bareness, without the context of a face, and the gap caused by any ambiguity is filled by the person who receives the communication.

A word that is used in internet communication and which has always stuck out for me is "heh." It's used in conversation in places like chatrooms and comment threads. When I read "heh," I read it as a cynical laugh, a sneer. When someone responds with "heh" to something I've said I place it along a spectrum that includes "whatever" and "fuck you." The Urban Dictionary, an online dictionary for slang words and phrases, has multiple entries for "heh," including: a half-assed laugh, a semi-cynical laugh, a filler word when conversation stalls, a filler word for when you don't know how to respond, a half-laugh for when something someone says isn't funny, a decent alternative to "lol." When I've asked other people online how they use it, some even say that they use it to indicate that something is funny. This slippage means that the sense conveyed by the person who writes it is never fully carried. The interaction lacks all of the contextual clues – tone of voice, facial expression – needed for meaning to be clearly conveyed.

This loss of meaning doesn't just happen with words, but also with pictures. Take the example of emoji. Emoji transform a sentence. A smiley face can indicate that you are happy, a winky face that you are joking and in cahoots, a sad face indicates displeasure, a grin for irony, a smiling face with tears of joy says that something is hilarious, a crazy face with sticking out tongue that something is whacky, and an angry face that you are, well, angry. Emoji are there to add nuance, giving a visual, linguist sign that enhances a sentence.

Emoji originated in Japan, where they were popularised on mobile phones, but now they are used worldwide. The name "emoji" comes from the Japanese 絵 (*e*–picture) and 文字 (*moji*–written character). Emoji evolved from emoticons (a portmanteau of "emotion" and "icon"), but are not quite the same. Emoticons are ways of conveying facial expression or posture through text, e.g. :-) or :-/. Emoji evolved from emoticons in East Asia where more elaborate emoticons were created, for example:

(")(-_-)(")
(an upset face with raised hands)

¯_(ツ)_/¯
(shrug)

(ノ °□°)ノ︵┻━┻
(flipping tables)

and its cousin
(ノ^_^)ノ┻━┻ ┯┯ ノ(^_^ノ)
(putting back tables)

Operating systems and other software providers, particularly mobile-phone platforms, started replacing these emoticons with images along with non-emoticon emoji.

I use emoji a lot. I love to add nuance to text, play with meaning, or just embellish it for the sake of decoration. They can also be a useful shorthand for saying something, allowing you to express yourself visually rather than doing it with words. I use the sunrise emoji to say "good morning," the sunset emoji for "goodnight," thumbs up to indicate "yes," thumbs up with smiley face to indicate "yes and I'm happy about it," person with folded hands for "thank you," and grinning face for a big cheesy grin. Before emoji, and still now, I used emoticons like :) and :(and their Japanese equivalents, ^_^, O_O, and >_<. I use emoji like most people, under the assumption that they carry the meaning I intend.

A study published in 2016 by Grouplens at the University of Minnesota shows that the implications of emoji use are more complex than they appear at face value. Emoji are interpreted differently depending on who is looking at the emoji, and depending on which platform they are using. The study looks specifically at how both semantic meaning and sentiment are communicated in anthropomorphic emoji (smiley faces and hand gestures, for example). It discovered that there is a wide range of interpretations for some emoji, both on the level of sentiment and of meaning.

For example, 44% of survey participants describe the sentiment for Microsoft's "smiling face with open mouth and tightly closed eyes" 😬 as negative, while 54% see it as positive. This varying degree of sentiment construal means that half of the people who read it won't understand the original intention of the person who wrote it. A sentence that you think has a positive spin could come across as negative. A single emoji can also have a different semantic interpretation: Apple's "unamused face" 😒 was described by different people as "disappointment," "depressing," "unimpressed," and "suspicious." An emoji like this makes your text more ambiguous, rather than less.

Emoji are displayed using a computing industry standard called unicode, set by the unicode consortium. Unicode was created to ensure that text is consistently encoded, represented, and handled across the internet. It ensures that developers have a universal way in which they can deal with all of the world's writing systems. A standard, unifying character system makes it possible to translate computer software. So U+0470 will render Ѱ in the Cyrillic alphabet, U+20A5 renders the ₥ currency symbol, and U+22C8 renders the ⋈ mathematical symbol. Emoji, like every other character set, is written in unicode. The unicode consortium provides a name for the emoji and a code. For example, U+1F602 is "face with tears of joy," U+1F618 is "face throwing a kiss," and U+1F64F is "person with folded hands."

The unicode consortium provides the code and the name, but it doesn't prescribe how emoji should look. It's up to each platform to render the image; what the emoji actually looks like depends on the platform that you are using. Anthropomorphic emoji may be simplified renditions of emotions, but it's up to the graphic artist to render those emotions. Each platform has its own graphic artists with their own subjective ways of interpreting emoji. Emoji look different whether you're looking at iOS, Google, Microsoft, Twitter, Facebook, or one of the many other platforms we use day to day. So the emoji that I send from my iPhone looks different when it arrives on my friend's Android, and the ones that I type into Facebook messenger look entirely different to the ones on my iOS native emoji keyboard.

Grouplens looked at platform-to-platform emoji interpretation and, unsurprisingly, found semantic meaning and sentiment are often lost in communication across platforms. 41% of the emoji they looked at had a different sentiment attached to them when looked at across platforms. The emoji "grinning face with smiling eyes" is interpreted much more negatively on Apple 😁 than it is on any other platform 😃 😄 😜 😄 (Google, Samsung, Microsoft, Facebook); and "sleeping face" is construed as neutral or positive on Samsung 😴 but as negative on Microsoft 😑 .

Beyond sentiment, there is a wide range in how a single emoji is interpreted semantically depending on how it is rendered. For example, "person raising hands in celebration" has the most variance in its interpretation: on Apple 🙌 it is described as "hand, celebrate"; on Google 🙌 as "stop, clap"; on LG 🙌 "praise, hand"; on Microsoft 🙌 "exciting, high"; and on Samsung 🙌 "exciting, happy."

The crucial points that the research uncovers are that emoji don't always carry the meaning intended by the sender, and that the variation in platforms that we all use mean that the platform itself can interfere with the delivery of meaning. This variation demonstrates just how implicated we are in the technologies we use. My choice of platform affects the meaning I convey. Who I am and the tool that I use are linguistically linked. The broad range of choice of tools and platforms means that even if we all speak English we no longer speak quite the same language. I send an emoji on my iPhone in celebration but the person receiving it might take it as an indicator for them to stop what they're doing, a cheery emoji sent from my Android looks aggressive on my friend's iPhone.

The complexity doesn't stop there: the Grouplens study was based in the United States, but many remote workers like myself work with people across different countries, from places with very different cultural signs and signifiers. A thesis on unconventional means of communication, which includes emoji, looks at how emoji are used in blog articles and comments. The research found that Japanese and American bloggers use emoji in very different ways. Japanese bloggers use many more emoji than their American counterparts. Their use reflects cultural themes such as *kawaii*, language play, and technology. They also go beyond carrying semantic or pragmatic meaning with emoji: they use emoji to appear more polite and create a more harmonious environment. The author found that Japanese bloggers use more emoji to make up for the lack of visual and auditory cues online, reflecting Japanese face-to-face communication which is heavily reliant on non-verbal cues.

When we come to a computer to write, to communicate, we come loaded with all of our cultural and linguistic baggage. It's not surprising that Japanese and American bloggers have a different approach to emoji. The wide use of English amongst remote workers makes it appear that we are all speaking the same language, but we rarely are, and it's easy for anyone to be misunderstood. Sarcasm, irony, and wit can come across as rudeness, sadness as standoffishness. There are people who have so much difficulty communicating online that they end up alienating themselves, without ever really understanding the reason why.

There are countless tools that facilitate online communication. It's easy to lose track of them all, to be unsure of where your next message is going to arrive from, whether it's iMessage, WhatsApp, Facebook Messenger, email, Skype, IRC, Telegram, Twitter, Facebook, Snapchat, GChat, Viber, WeChat, SMS, Instagram, or Slack. There is a constant barrage of incoming messages to be fended off.

Usually we take these tools at face value: they are simply a way to message one another, a way of staying in touch. But they are not just tools. Each media creates the conditions for communication; it sets the rules for the way that you and I interact with one another. We lose the immediacy of human face-to-face communication and instead what we say and what we feel is mediated through a set of decisions that were made by someone else. Communication media are not just tools–they are a form of architecture. A communication medium creates a social space where communities form. It creates a virtual coffee house, bar, classroom, park, playground, and office. Beyond the content of the message, what matters is its form, whether that is the instantaneousness of a direct message, the pause of a blog comment, or the 140 characters of a tweet. Our content takes the form given by the medium's architects.

An obvious example of this is Facebook's Like button. Introduced in 2009, the Like button provided Facebook users with a way of expressing their liking for a person's post: someone can post their cute cat's picture and someone else can click Like, and everyone feels good about it. It is a frictionless way to participate in someone's content, and someone's life, without having to use words. It was followed by rumblings on the internet: what about a Dislike button? If I can Like someone's content, shouldn't I also be able to Dislike it? It felt like common sense that if one is able to express sentiments on a platform, they should be both positive and negative. But Facebook had no plans to introduce a Dislike button. In a 2014 live Q&A, Mark Zuckerberg was upfront about Facebook's reasons for not including one. He says, "Some people have asked for a Dislike button because they want to say, 'That thing isn't good.' And that's not something that we think is good for the world so we're not going to build that."

The architects of Facebook don't think a Dislike button is good for the world. Which is understandable; they are creating a platform that they want people to return to, a place where they feel supported and bolstered. If a person experiences negativity why would they return? It's one of the key insights of behavioural psychology. To get someone to act how you want, give them a positive reward. Negativity does not have the same effect. Instead, in 2016, as a way of allowing people to better express empathy, Facebook introduced "Reactions," allowing users to express the emotions of Like, Love, Haha, Wow, Sad, and Angry. Like all design that looks simple, Facebook Reactions went through an intensive design process, driven by analysis of the huge amount of data gathered by Facebook, a process that encompassed research into cross-cultural image interpretation, non-verbal communication, and user-interface design.

Facebook Reactions, and the Like button, are forms of language. They are a way of communicating sense and meaning to another person. David Smail writes, "Whoever controls language, controls thought." In constraining the way a person can react to a post to positive emotions, Facebook has created a space that is marked by its positivity. And the thing that keeps us coming back is the social approval that we all crave.

This communication platform, which, as of March 2017, is actively used by 1.94 billion users, does not just straightforwardly provide a way for people to keep in touch, share images, and communicate. By determining the way that people can react to a piece of content, Facebook manipulates and controls a popular social space. Control of the form of our social interactions is handed to a small group of people who decide what is and isn't "good for the world." It makes sense for Facebook to create a space that is positive, that people want to return to, because then advertisers want to use the platform and the company can increase its revenue. However, is that necessarily good for our relationships, or, if we think in terms other than good or bad, does it create a space where the full richness of human relationships can be experienced? How do we go on to form and maintain friendships in a world that is constrained in such a way? How can we deal with the more difficult aspects of relationships if we spend our time in a mode that is only ever positive?

Alongside the tools for social interaction, the communication tools that I use for work have evolved too. As a remote worker, these tools provide the fabric and background of my online working life. They are my main substitution for an office.

Every tool sets different implicit rules for communicating. Email is one of the most established online communication tools, and still the most popular, especially in business. Email allows a person to send a message to a server which is then delivered to a person once they are online. When I send an email to someone I expect a response, but I usually don't expect an instantaneous response. Perhaps I send a message and hope to get something by the next day. An email thread sets up a back-and-forth response system in which it's expected that the person forming the response will need time to do so. Instant messaging services are different. Because of their brevity and the speed with which they can be sent, a message sent elicits an almost immediate response.

Many distributed teams use chat clients to create a virtual office. Rooms can be created where team members can interact with one another, socialise, and work. However, these clients can insinuate themselves into your day-to-day life. Today, a tool that is ubiquitous among remote teams is Slack. At its most basic Slack is a chat platform. Prior to using it I used a mixture of Skype, IRC (Internet Relay Chat), and email. However, Slack has subsumed all of these communication tools and become the main place that I communicate for work. It is as close to an office as a virtual worker can get.

Slack's growth has been impressive. From its launch in August 2013 until April 2016, it has acquired 2.7 million daily active users, with 800,000 paid seats. Unlike many tech platforms, these aren't just people who sign up and forget about it. The average Slack user spends ten hours per day using the platform.

Like these average Slack users, I spend a lot of time on it too. There is a lot to love about it: I am part of a team of people working together in a carved-out virtual space on the internet; it makes the team feel like more of a team, all in one place, together; it makes communication between team members easy on both desktop and mobile. Unlike social-media platforms, Slack doesn't generate its money through advertising, but through signups. This means it has an interest in making the product indispensable enough that you will pay for additional features, making it err on the side of useful rather than just plain addictive. However, there are lots of tropes that Slack makes use of as much as any other platform: the urgency of red notification symbols places a demand on users to respond to messages; the continuous nature of chat creates a fear of missing out (FOMO) which keeps people on the platform; the message you get when you set up a new team suggests that you turn on notifications.

Because communication is instantaneous, it's easy to feel like I always have to respond instantly to a message. In an article outlining the pros and cons of group chat as a primary communication method, Jason Fried, CEO of Basecamp, describes chat clients like Slack as creating an ASAP culture, one in which people feel that things need to be done straight away, right now. Fried goes on:

> Most things worth discussing at length are worth discussing in detail over time. Because chat is presented one line at a time, complete thoughts have to unfold one

line-at-a-time. But since people can jump in any time before you've had a chance to fully present yourself, making a point can become really frustrating really quickly. Further, incomplete thoughts and staccato responses make it really difficult to fully consider a topic and make important decisions – especially in a group setting.

There are other problems with chat clients: their continuous nature means that you always feel like you are trying to catch up with what's going on, it's easy for decisions to get lost if they aren't also documented elsewhere, and distributed teams end up in situations in which they are always online. When you work with someone in Australia and they send you a message late at night it's hard not to feel obliged to send a response to that person. There are many ways to mitigate against these problems. A company can create communication guidelines which let people know what is expected and individuals themselves need to be equipped to properly manage such high-frequency communication. When people start working online they can quickly feel overwhelmed, unable to cope with the high volume of information coming at them. All new remote workers have to quickly develop new skills that enable them to deal with this unfamiliar way of working. By the time their second year starts many have learned the techniques they need to cope with so much communication.

All of our communication is mediated by online platforms that shape us to behave in particular ways, but it's likely that none of the internet's original architects, nor those who build social and communication platforms, would have anticipated the emergence of trolling. Trolling is at the extreme end of the internet's communication challenges. It is not just aberrant or a phenomenon that occurs in isolation. Trolling is along the spectrum of many of the communication challenges I've already discussed; it happens at the convergence of human behaviour and technology. It's hard not to wonder if there is something about the medium that brings out the very worst in all of us.

In Plato's *Republic*, Glaucon, Plato's brother, asks whether any man would be so virtuous as to resist temptation when he knows that he will not be discovered. He says that people only act with morality to maintain a reputation for virtue and justice. He tells Socrates the story of Gyges, a shepherd who finds a ring that, when he is wearing it, gives him the power of invisibility. He uses the ring to seduce the queen, murder the king of Lydia, and become the king himself. Glaucon uses the story to make the point that morality is a social construction and when human beings are given the cloak of invisibility and anonymity, away from the gaze of other people, they will act unfettered by morality and in the most terrible ways.

Internet communication is marked by the absence of the gaze. Like Gyges, when we interact online we are given invisibility and have the opportunity to act outside of morality. For some people, this anonymity is license to act in ways they wouldn't normally if they were subject to a mediating gaze. This isn't an unusual phenomenon. A 1981 study of suicide baiting found that in ten of twenty-one cases of people threatening to jump off a bridge, building, or tower, people in the surrounding crowd shouted for the person to jump. The researcher identified various common de-individuating factors across the ten cases, including that they were members of a large crowd, there was a physical distance between the crowd and the victim, and the cover of nighttime. There has been ongoing research about how the presence of a human gaze alters our behaviour: a study at Newcastle University found that when images of eyes were displayed on the walls of a cafeteria the incidence of littering was halved; a 2007 study found that people who are "watched" by a robot with human eyes are 29% more likely to act in the common good.

All animals have an involuntary system called "gaze detection" which detects and reacts to the presence of a set of eyes. This neural architecture means that we are highly sensitive to being looked at–not only is this automatic but it is impossible for us to turn off. Some researchers have concluded that gaze perception is hardwired into us. We are so sensitive to the gaze of others–what happens when we remove it? That depends on the structure of the society or network that we are caught up in. To return to Foucault's metaphor of the Panopticon–the circular prison with the central tower which places all inmates under a constant state of surveillance–there are systems in which behaviour can be controlled just with the knowledge that we might be being watched. Even the possibility of the gaze can control behaviour.

But the internet removes the gaze entirely, providing the illusion of anonymity. This leads to what's called the "online disinhibition effect": the reduction of social constraints that would be in place due to the presence of other people. When it is benign, the online disinhibition effect leads to people sharing more of themselves and breaking down barriers. This can have a very positive impact for some people, allowing them to connect more deeply with others and perhaps share experiences of pain or trauma that they wouldn't be comfortable sharing in person. When it is malignant, it is called the "toxic disinhibition effect." In both cases, anonymity, invisibility, and particularly lack of eye content, lead to disinhibition which allows people to act in ways they wouldn't if they were restricted by the presence of a gaze. John Suler, who coined the term, has identified six factors that contribute to it: 1) *you don't know me* – most of what I do is anonymous and therefore not linked to the rest of my life so I don't have to take responsibility; 2) *you can't see me* – I cannot see a physical reaction, such as a frown, tear, or smile, which might inhibit my action, and the person I'm interacting with cannot see my own physical reactions; 3) *see you later* – online communication is asynchronous so I can say something, disappear, and return when I am ready for the response; 4) *it's all in my head* – because I read text and hear the other person's voice in my head, I internalise them and place them within my own intrapsychic world, my imaginative realm, treating them more like a character in a book than an actual person; 5) *it's just a game* – some people see online life as just a game that has different rules and norms to their everyday lives, so that when they leave the game behind they leave behind their online identity; 6) *we're equals* – online we all lack indicators of status and authority so we are more likely to speak out against injustices or misbehave.

At the most malignant end of the online disinhibition spectrum are cyberbullying and trolling. The internet is rife with trolls, people who are doing it for the "lulz," putting their own enjoyment ahead of the feelings of others. A 2014 research study found that people who self-identify as trolls are more likely to display high levels of the Dark Tetrad personality traits–narcissism, Machiavellianism, psychopathy, and, particularly, sadism. Trolls enjoy the pain and discomfort of others and the internet affords them infinite possibility for that pleasure, with no repercussions. The authors conclude that trolls are just opportunistic sadists who are acting on their instincts. In this view, the online disinhibition effect reveals the true self that society forces us to hide.

However, that is a rather simplistic explanation of trolling, one that ends with the shoulder-shrug of "well people are just people, and some people are bad." A later study carried out by Stanford and Cornell researchers found that in the right circumstances anyone can display trolling behaviours. Factors such as being in a bad mood, or even just seeing a trolling post from another person, can lead a person to themselves act like a troll. In the study, people in a good mood who saw no trolling comment on an article left trolling comments 35% of the time; if they were in a bad mood or saw a trolling comment this jumped to 50%; but if they were both in a bad mood and they saw a trolling comment they trolled 68% of the time. This presents a more complicated picture than the simplistic view that trolls are just anti-social sadists. Even if we accept that some trolls are out to enjoy themselves, their presence and the messages that they post contribute to escalating negativity, creating trolling behaviour in people who don't normally display sadistic personality traits. Anyone can be a troll.

The environments in which we communicate mediate how we act. The medium is the message. While there have always been indications of the disinhibiting effects of anonymity and invisibility, it's only in the past twenty years that we've seen the impact that it has on a global scale. Everyone can be anonymous. Everyone can act in ways that they wouldn't if there was a set of eyes upon them. Some people choose to take full advantage of this and spend their time trolling others, and there are people who troll just occasionally, because they are in a bad mood or because they are spurred on to do it. For me, it sometimes takes the physical removal of myself from my computer to prevent me from saying something cutting or hurtful to another person.

Online, we operate and oscillate between two poles: on the one hand, there is the disinhibition created by the internet which allows people to either share more of themselves or to act in ways that they wouldn't normally, and on the other, there is the necessity of the performance, of creating a self as we wish others to see us. Both of these poles are made possible by the internet itself and the platforms on which we communicate. They form an aspect of our identity which was absent before we were so hyperconnected, before we were given such high levels of anonymity, before we had this hystericised feeling of performing for others. Is this an inevitable result of the creation of a global communications network, or could an earlier intervention have allowed for more of a direct connection with other people? Is that intervention even possible anymore?

It's ridiculous to talk of some sort of authentic, unmediated experience, one in which we completely eschew the internet and return to an Amish-like state of nature. The internet is a fact of our lives and we are all implicated in it. We are all part of the network. We are not, and have not been for millennia, natural beings, we are technological beings; technology isn't just a tool, it forms the fabric of our environment. It creates, effects, and shapes our relationships, and we are living through a paradigmatic shift in which escalating changes in technology are having a tangible impact on who we are and how we relate. The question is not one of how we go back, but how we move forward. When we communicate online, the gaze of another person, that key component in the creation of empathy, is absent. How do we recreate empathy in our online spaces? Or are we already beyond it? Is empathy itself an anachronism? Are we living in a world beyond empathy, where what matters most is the communication of information?

Maybe if we look to Socrates we can find a fleeting moment of instruction. After hearing the story of Gyges, Socrates responds that the shepherd's transgressions are not because he is uninhibited by the social construct. He says that the man who uses the ring is enslaved to his appetites, while a man who chooses not to use it remains rational, in control of himself, and happy. It's as though Socrates speaks to us across millennia: we all wear the ring of Gyges, in thrall to our appetites, consuming information, acting as though no one can see us, but if we want to hold on to our humanity we need to maintain a cautious distance, take the ring off, remain skeptical. Rationality and inquiry lead us to happiness, not the mindless pursuit of easy pleasures.

CHAPTER EIGHT
DESIGN MY LIFE

We were sat in a minibus, driving along the busy route from Mai Rim to Chiang Mai. The bus was a shuttle provided by our resort; clean, sterile, air-conditioned. Outside, the traffic was in chaos. Cars of all descriptions, pickup trucks packed full of Thais, their mouths protected from the pollution by grubby white handkerchiefs. Scooters bobbed and weaved through the traffic, each a hairsbreadth away from a major collision. Many of the scooters carried not just one person but whole families, with parents and children perched precariously together, hurtling through the traffic.

We were in Chiang Mai for two weeks of our year-long travels. We spent most of our three months in Thailand in Krabi, the southern province famous as the location of Phi Phi Island and Phuket, both magnets for travellers, backpackers, and scuba divers. We lived in a new apartment block on Klong Muang Beach, just a few miles outside the town of Ao Nang. During the early morning I would sit on the balcony, looking out over the pool at the emerald sea just a few hundred meters away, tapping on my keyboard, beavering away, waiting for D to wake up so we could start our day in Thailand together.

We were digital nomads: or, at least, I was a digital nomad. D was just a tagalong, having taken a year-long sabbatical from his academic job so that we could travel together, me working, him exploring the world one hammock at a time. When you work online, you are essentially location-independent, or location-agnostic, or whatever term you prefer to use to say that it doesn't matter where you work. Before I worked online I was tethered to an office, having to commute in for the day-to-day drudgery, only able to go abroad and travel during my meagre holiday allowance. Four years after I took my work online, D and I decided that it was time to take full advantage of my location-independent status. He was weary of his job and I was weary of England. Our relationship had improved a little since I had started a job and I was now able to keep my work time within certain boundaries. But I was still travelling all the time, having experiences without him, lonely when I was at home, and always on the verge of picking up my phone to check my email. My job paid enough to sustain us both (especially in countries where the economic conditions meant that we were essentially rich), so we decided to sell as much of our stuff as possible, put the rest in storage, and go on the road.

From about six months before we left we started selling things. We put all of our books of any value on Amazon, our furniture, computer, and television were listed on eBay and Gumtree. There was an endless troupe of people coming through our house to take our stuff: our television, our furniture, games consoles, and a never-used cross-trainer. We would take it in turns to go to the post office with stacks of books parcelled up in brown jiffy envelopes, causing a long queue of people giving us evil glances as each parcel was weighed and stamped.

I found it cathartic to sell so much of my stuff, all of the shit that I'd accumulated over the years. It's like shedding a huge amount of weight, stripping back to the bare essentials. However, it would be disingenuous to say that we sold absolutely everything. With the possibility of coming back, we kept hold of those things that we couldn't bear to part with–our most precious books (which was a lot of them), our bed, and an Austin Osman Spare painting we bought for ourselves as a wedding gift–and those things that we didn't really know how to get rid of–a suitcase full of my shoes, boxes of clothes that might come in useful again, paperwork, D's school books, a box full of cables, miscellaneous furniture that we hadn't been able to sell, photographs, memories.

I became a digital nomad, living by my laptop, my Wi-Fi connection, and my airline rewards programme. By the time we arrived home we had circumnavigated the globe: Netherlands -> Lisbon -> Thailand -> India -> Thailand -> New Zealand -> USA -> Belize -> USA -> Canada -> Mexico -> USA -> home. I did what was expected of me: photographs of my laptop on the beach parked beside a fresh young coconut with a straw sticking out, the strip of white sand, the bright blue of the sea, posted to Facebook with a huge sense of glee, laughing at all those working stiffs back home in the English rain. Of course, the reality is that I rarely worked on the beach. Sand does not play well with laptops. In Lisbon I worked from our chintzy Airbnb apartment, at Klong Muang Beach from a balcony on which I would get baked in the direct heat of the afternoon sun, in Caye Caulker at a cafe where a manky, bandana-wearing dog played around my feet. I worked from a string of cafes and a string of apartments, from hotel lobbies and hotel bedrooms, in airports and on trains, from the lobby of a cinema, on airplanes, in internet cafes. I was constantly hungry for good internet, the quality of which varied from country to country. Thailand had excellent connectivity; cheap 4G SIM cards and a huge number of internet cafes meant it was always easy to work. In New Zealand the internet was terrible. I moved from cafe to cafe, shunted ever onwards by the thirty-minute restriction that was in force on every Wi-Fi network, eventually buying a Wi-Fi dongle on which the internet chugged along. The USA fluctuated between high-speed tech hubs and desperately trying to squeeze a few megabytes out of hotel Wi-Fi. Whenever I arrived in a new country the first thing I would do was buy a SIM and find out where the best Wi-Fi was. I needed to be connected; if I was not connected I could not work.

That year of travel was filled with overwhelming experiences and is marked by fond memories, and yet it was fraught with questions about the nature of what it is to travel and about the way the internet has permeated not just our individual lives but the world as a whole. Travel has traditionally been the quest for absolute difference, a quest for complete immersion in another culture. When you take your work with you, you travel in a different way. It's hard to escape the ever-presence of the internet. By always being there it changes our relationship to the places that we visit. Wherever you go, it's always there. Of course, it's still a wonderful experience but a digital nomad will never experience the Patagonia of Bruce Chatwin, the Shenandoah of Bill Bryson, or the Turkey of Freya Stark. I had travelled a lot previously, with different levels of immersion: a gap year as a snowboarding bum in Whistler, traipsing through Palaeolithic caves in the south of France, following Alfred Hitchcock's footsteps through San Francisco, visiting archaeological sites in Turkey, months in Prague as an Erasmus student. All travel is done through a lens, whether you're a tourist, digital nomad, backpacker, businessperson, or student.

Despite being for many an escape, travelling as a digital nomad means experiencing a new place as a digital nomad, which gets in the way of being fully immersed within a culture. There is always a layer of reality that is the same, wherever you go. The internet has changed our very relation to place. This is the thought that struck me as we sat on that minibus travelling from Mai Rim to Chiang Mai. As I stared out the window, at a world so different to the one that I knew, I found myself thinking about work, thinking about the same problems and concerns that I had been thinking about six months previously, while sat in my office in Canterbury. That morning I had got up, switched on my laptop, and talked to the same people I always speak to. I had checked my email, talked on Skype, and done some work. Things were different, and yet they were still the same. My location had changed, but my thoughts had not; they were just variation on the same thoughts I had at home.

Staring out of that minibus window I experienced a sense of vertigo, an overwhelming feeling of the totality of the internet. A seething vastness of voices, of people and opinions, a web of connections, an infinite, undifferentiated multitude that contains the totality of the world. A layer of reality. Reality itself. The internet as crawling chaos, which is there everywhere you go, always the same, always filled with the possibility of difference.

It made me feel small. It made me feel lost, as though I was looking over a precipice into some vast sea. The internet is a real and virtual place, self-generated by us and yet in its totality taking on an identity of its own. Immanent to us but separate, having taken on an element of transcendence. I realised how utterly this virtual layer of reality had transformed my experience of place. That it is vast, and it is homogenous, and it will always be there, wherever I go, always the same, ready to be jumped into. I wonder if it will ever be possible to experience the world without the internet ever again.

Over the past ten years the term "digital nomad" has become increasingly familiar. It's not a new idea: rich people zipping around the world, untethered to their location. In 1990 Jacques Atalli, banker and advisor to the French President, published a book in which he talks about "nomadic man," describing the computer as the "nomadic man's passport." A later book in the 1990s, written by Tsugio Makimoto and David Manners, bears the title *Digital Nomad*, and explores the travel possibilities that our technologies will bring us.

As laptops became more affordable and the internet more ubiquitous, the conditions became right for digital nomadism to emerge as a global phenomenon. Today's digital nomad, however, is different to the jet-setting elites that were predicted. A digital nomad is not necessarily rich. The economic conditions for young people and those in their thirties and forties in their home countries are increasingly tough. In 2012, about half of all US college graduates under the age of twenty-five were out of work or underemployed. In 2015, the UK government reported that around 16% of graduates between twenty-one and thirty were unemployed, with 31% of those who did have employment doing medium- to low-skilled work. Those who are employed find their wages stagnating as the cost of living increases and prices inflate. Many young people find themselves unable to afford to buy houses, or even, in places like London or San Francisco, afford extortionate rental costs.

Within these challenging economic conditions, there is some hope. Many young people are highly computer-literate, and with sites like Upwork and oDesk, there is a very low barrier to entry to find work online. There are also multiple opportunities for setting up your own business, working as a freelancer, or getting a job at a remote company. If all you need is a laptop and an internet connection to work, why not go somewhere where the cost of living is low enough for you to enjoy a higher quality of life than you could at home?

Digital nomads are often freelancers or early-stage entrepreneurs, who are seeking a better quality of life than the one that they can afford in their home country.

Take, for example, Pieter Levels, a digital nomad who was profiled in WIRED in 2014. On a mission to travel the world while launching twelve startups in twelve months, he sold everything he owned and set off for Thailand, in search of a different way of life and a lower cost of living. The internet is full of stories of people doing the same thing, getting rid of everything and heading off for something different.

Then there are those like D and I; people in their thirties or forties who are tired of life in their own country, frustrated by the little they can do with their money, unable to get on the housing ladder, perhaps sick of the weather, possibly sick of their careers, who leave home in search of a better quality of life.

Research conducted by Harvard's Desirability Lab at Ubud in Bali found some notable features in digital nomads: the majority are freelancers or run their own small business (75%), many of whom have set up low-cost businesses at home before going on the road. They usually earn money in jobs that are flexible, particularly in the tech industry, so design, development, marketing, but also SEO, online sales, and life coaching. While 34% of those on the road were millennials, the largest group was mid-career people, who had often left established corporate careers that they didn't enjoy in search of a different way of life. "When we asked what prompted the choice to go nomadic, the specific reasons differed, but the arc was strikingly similar. They had not enjoyed their work for a long time, and a crisis–of identity, or relationship, or change of circumstance–nudged them to make a major change." The implication here is that for a large number of people, the way of life that focuses on career and money leaves them lacking something important; call it meaning, or existential satisfaction, or whatever. The internet enables people who are trapped in their jobs or their lives to more easily go in search of something else.

Digital nomads come in all varieties. Beyond freelancers there are people like me who work for distributed companies that allow their staff to be location-independent. There are entrepreneurs who set up lifestyle businesses to sustain themselves, or with loftier goals of getting investment and making it big. Some make money through their nomadism, through writing or life coaching. Then there are those who have already made it big, financiers and successful serial entrepreneurs who dress in the same shorts and flip-flops but whose $25,000–$50,000-per-month income place them, in countries like Thailand and Mexico, amongst the super-wealthy. Other high-earning nomads include individuals who have been early employees at successful tech companies, and elite freelance developers who attract high salaries.

There are also digital nomads who get by through coaching and helping other people to become digital nomads. A new industry has grown up with technologies and programmes that are aimed at digital nomads: for example, there is Nomad List, which is a list of the best places for digital nomads to live ranked by cost of living, internet connectivity, and weather; there is the Digital Nomad Academy, which promises to teach you how to pursue an alternative lifestyle, selling itself as "a place for ambitious and hungry upstarts who want to participate in the creation of an Entrepreneurship Economy, and push the envelope as we help enable each other, and the people around us, to realise a new way of living"; and there are the countless blogs that sell the digital-nomad lifestyle, making money through consulting, advice, and advertising.

Much like the entrepreneur, the digital nomad is an archetype who has been able to emerge through the convergence of technology and ideology, living many of the neoliberal ideals of freedom and liberty. They are independent, entrepreneurial, and act in ways that best suit themselves; not only are they untethered from institutions, but they are untethered from place. They can be anywhere, do anything, be anyone... provided they have a good internet connection. They may not define success in terms of money or power, but they define it in terms of the creation of the perfect lifestyle.

As a phenomenon, digital nomadism has only become possible in the past fifteen years. If, when I had left school, I had been able to hustle for work online and live anywhere, I would have gotten as far from the UK as possible. The internet is facilitating a transition into an age when space does not matter. Many of us can be anywhere and have access to the same opportunities that we have at home. This isn't just in the area of work, but in access to education and information. We can travel, work, and learn anywhere, individually reaping the benefits of the global network. For kids growing up now, this won't be new or exciting, it will just be a fact of life, and as I watch these changes happen I wonder if the impact of digital nomadism will just be a positive one.

Though a study is currently underway at Chiang Mai University, there haven't yet been any substantial studies of the impact of digital nomadism on local communities. Therefore it's hard to tell what the potential long-term impact will be on these communities. We can, however, looks for clues in similar transient groups of people. Tourism is one of the world's largest industries, with a global economic impact of around $7.6 trillion in 2014. Tourism is generally seen as a positive for a country: it brings investment, creates jobs, promotes sales of local artefacts and crafts, and increases global awareness of cultural traditions.

Large groups of tourists have always had an impact on host nations. Cultural heritage sites lose their socio-cultural value, instead becoming economic commodities. Such changes aren't just caused by mass tourism: the interest of wealthy English grand tourists in antiques, particularly portable antiques such as pots, coins and works of art, led to price inflation of such items. By the eighteenth century there was a widespread black market, with objects of cultural heritage clandestinely removed from archaeological sites and sold to young, wealthy aristocrats. The convergence of large groups of people on cultural heritage sites can lead to an increase in pollution, damage to the site, and even vandalism as tourists seek to collect mementoes. Examples of this are numerous, from the Finnish tourist who chipped the earlobe off an ancient Moai on Easter Island, to two Californians who carved their initials into the wall of the Coliseum and took a selfie with it.

It's possible to trace the roots of the digital nomad back through subsets of the tourism industry. There is a long tradition of young people, particularly

Westerners, travelling to foreign countries, to experience new cultures, to find themselves, to expose themselves to artwork and music and different ways of life. From around 1660, wealthy gentlemen embarked on a "grand tour," taking in the sights of Europe in an attempt at cultural betterment. Today's digital-nomad movement has much more in common with those young, upper-class men, and later women, than it has with nomads. The grand tourists travelled across Europe, meeting with people in the same social class and other grand tourists, but largely staying on a preset path. The aim of a grand tour was personal betterment: it was thought that exposure to works of art, pieces of music, architecture, and the aristocracy of different cultures gave the tourists a level of refinement beyond their peers who stayed at home. This type of cultural enrichment was essential for any young gentleman. In the words of Samuel Johnson, "A man who has not been in Italy is always conscious of an inferiority, from his not having seen what it is expected a man should see."

The establishment of the railway system and steamship meant that travel became more affordable and opened up tourism beyond the aristocracy, growing in popularity particularly in the middle classes. While the notion of the aristocratic grand tour died out, the twentieth century saw an explosion in other groups of people going in search of themselves: from the 1950s beatniks, to the 1960s and '70s when hippies travelled east to find themselves, to the growth of gap years, travel has been a way of refining yourself, broadening your perspective, and changing your worldview. It is in this tradition we find digital nomads. They are the backpackers of the digital age. By looking at the impact of backpackers we can see some of the ways that digital nomads may have an ongoing impact on local communities and economies, and we can also make predictions based on what differentiates them from the backpacking community.

Backpackers are distinct from other tourists in a number of significant ways: they spend longer in one place, travel for extended periods, are usually on a budget, and they seek out local and authentic experiences. Various studies have shown backpackers in a favourable light compared to other types of tourist. There are a number of reasons for this. For one, they stay longer, spending more money within the local community. This means that there is less economic leakage (the process whereby tourism-related revenue is lost to other countries). In normal tourism, leakage is around 70%, whereas in backpacker tourism it is around 30%. While tourists often demand imported and familiar brands, backpackers usually favour local goods which tend to be cheaper and give them more of the authentic experience they are seeking. The employment opportunities for locals who support the backpacker infrastructure are qualitatively better than those generated by the tourism industry (though quantitatively less). While the mass tourism industry may generate jobs like being a chambermaid in an international chain hotel or being a waiter in a hotel bar, the backpacking industry creates jobs such as running a guest house or home stay, becoming a dive guide, or leading treks. And, as one of the main selling points that a backpacker is looking for is an authentic experience with a local, setting up your own backpacker-targeted business has a lower barrier to entry. Budget-minded travellers usually expect things to be rough around the edges. Beyond that, backpackers tend to be more respectful of the local community, seek meaningful relationships with locals, and are friendlier and more sensitive towards the local environment.

The fact that backpackers, and if we want to draw parallels, digital nomads, stay longer in one place does not just have positive benefits. The longer degree of contact can mean that they are more invasive into local cultures, customs, and spaces, including homes, ceremonies, and sacred spaces. Backpackers, like most tourists, are focused on having fun and extracting as much as possible from the time they have available in a place. A study in southern Thailand found that the hedonistic attitude of backpackers, which includes parties, drugs, public sexual conduct, improper dress, and binge drinking, leads to resentment among locals who see their customs and traditions being disrespected and destroyed. Similar studies have been carried out across popular backpacker areas including Goa in India, Gili Trawangan in Indonesia, and Zipolite in Mexico.

A 2016 paper looked not just at the perceptions of backpackers by locals but at their mutual perceptions, focusing specifically on backpackers in the Yasawa Islands in Fiji. The study found that where locals saw the negative impacts of backpackers, the backpackers saw themselves as neutral. For example, locals saw disrespect towards their beliefs, including scanty dress when visiting villages or religious places. The relationship between backpackers and locals caused an increase in the consumption of alcohol and drugs amongst locals, with backpackers bringing marijuana to the island and increasing local demand. There have also been changes to the Fijian way of life, with an increased reliance on imported and tinned food, changes in modes of dress that have been influenced by visitors, a decrease in the number of people attending church or religious ceremonies, a decrease in respect to elders, and a decrease in family unity and family values. On the flip side, locals and backpackers agreed that the influx of backpackers has helped to revitalise customs. One Fijian commented that "Backpacker tourism helps in preserving *Bure* [traditional Fijian house] as we are trying to build more *Bure* in the village in order to educate tourists." There has also been a growth of intercultural awareness as backpackers and locals interact with one another.

Overall, local Fijians are much more sensitive to these changes and perceive them as negative, whereas backpackers, even if they are aware of any possible impact, remain neutral. It's not surprising that hosts find their society and environment transformed by the transient community. Backpackers and other tourists, while they may have no negative intentions towards host communities, are generally focused on themselves, looking for fun or self-discovery, new experiences or self-betterment. They go somewhere to achieve this aim of developing their self; the communities that host them are primarily a vehicle for this.

Digital nomads are a distinct subset of tourism and have many similar features to backpackers. They usually travel with a rucksack to several destinations, taking longer trips than conventional tourists, and have flexible itineraries. Depending on the income they generate, they may or may not use budget transport and accommodation, and may or may not be budget-minded. While the stereotypical backpacker draws out their pennies by drinking cheap local drinks, the stereotypical digital nomad will pay whatever they have to for the perfect latté.

Like backpackers, digital nomads often avoid typical tourist destinations, looking for places off the beaten path and searching for an experience that is "authentic." They do, however, create and flock to their own hubs. Backpacker hubs, like the Khao San Road in Bangkok, are defined by a high number of hostels, restaurants and cafes catering to backpackers. Digital-nomad hubs, like Chiang Mai in Thailand, Berlin in Germany, Udub in Bali, and Playa del Carmen in Mexico, create their own unique zones, defined by co-working spaces and cafes that cater to the tech worker's constant desire for high-speed internet and high-quality coffee. The irony of many backpacker and digital-nomad hubs is that they create mediated zones where an individual can experience the difference of a new culture while still having many of the same comforts as home. Familiarity lies in internet cafes, hostels, coffee shops, and restaurants, and difference lies in the exotic other who is always just a few feet away. You can always venture out safely knowing that there is something familiar to return to.

From the similarities and differences between digital nomads and backpackers, we can make some predictions about the impact of digital nomads. The most notable similarity is that they spend longer periods of time in one place than a regular tourist. This could be anything from weeks, to months, to years.

While there is undoubtedly some economic leakage, there is likely to be less than with a regular tourist.

Like backpackers, digital nomads are often in search of authentic experiences. They are more likely to spend money with locals than regular tourists, supporting locals and contributing to cross-cultural exchange. However, because of their relatively high income levels, they don't necessarily need to spend money with locals, and can afford to pay for home comforts such as imported goods or Western-targeted products and services with a high markup.

The authentic experience sought by digital nomads is, of necessity, more highly mediated than the one sought by backpackers. Digital nomads need to have zones where they can work, which means that they need to be in places with adequate infrastructure (though often a 4G SIM card can let you go pretty far off the beaten path). Also, as I experienced myself so vividly on the road to Chiang Mai, the fact of working on the internet prevents you from ever being 100%, or even 50%, immersed within a culture. Wherever you go, there is a layer of reality that does not change: you could be trekking through the Amazon, sitting on a beach in Australia, visiting a temple in Cambodia, taking a train through India, and you find yourself thinking and worrying about exactly the same things that were on your mind at home. The only thing that has changed is the backdrop. When you take your work with you, this level of homogeneity is inescapable.

Digital nomads usually have more money than backpackers. They may have saved some money before they embarked on their travels, but they often have a regular and stable income. As many digital-nomad hubs are in less developed countries where the cost of living is low, digital nomads often have a high level of disposable income. Whereas a backpacker might need to keep their spending below $15 per day, a digital nomad is unlikely to have such restrictions and will have more money to spend in their host country.

Backpackers travel on a budget; this means cheap train travel, coaches, and, increasingly in recent years, low-cost airlines. Digital nomads don't necessarily have to travel on a budget, and while they often look for bargain travel they are able to travel more luxuriously should they need to. Often they are members of airline or hotel points schemes, which they collect religiously, sometimes travelling on specific routes to gain status points and therefore get access to perks like airline lounges and upgrades.

A major distinction between backpackers and digital nomads is that digital nomads often have ongoing responsibilities which mean that they can't live as hedonistic a lifestyle as a backpacker. Like most of the tech industry, they party, but their parties may be tempered by the fact of getting up to work the next day. They often have clients to service, regular employment, or a business to run, so they tend to need some regularity to their lives.

Digital nomads also need to factor in the impact of travel time and location changes on their work. I found my work productivity dip considerably when I had to travel from place to place. There isn't just the travel time, but the mental effort it takes to re-establish a working pattern in a new place and often a new time zone. I found that I worked best when I was in one place for an extended period of time, where I could have a routine that allowed me to spend long portions of the day working.

A significant difference between digital nomads and backpackers is their means of generating income. Backpackers, when they need money, do seasonal work or casual labour. They are likely to have arranged a work visa for the country they are staying in, so that they can raise money if they need to. They might work for a few months at a resort, work as a nanny or au pair, in a bar or restaurant, participate in a WWOOFing programme (where they do work in exchange for food and accommodation), teach English, or do fruit or vegetable picking. The benefit to being a digital nomad is that you don't have to worry about finding work or money while you are travelling—you take your work with you. This means, however, that you are generating income and paying taxes in another country while benefitting from the infrastructure in your host country. When I was travelling and working, I paid taxes in the UK, so while I did spend money in my host countries, I didn't contribute to the economy via taxation. The country I was in was supporting me, but I wasn't supporting it. As more backpackers make the transition into doing digital work, there will undoubtedly be negative effects experienced by those types of businesses that normally rely on their labour.

Digital nomads often travel to places where it is relatively easy to get a visa. Because they aren't legally employed in their host country this usually means a tourist visa. The length of time they spend in a host country may be determined by the length of their visa, then they travel on to countries in the visa waiver programme or where it is quick and easy to acquire a tourist visa. In countries like Thailand which have fairly lax visa restrictions, digital nomads can extend their stay with a quick hop across the border to get a new one. As a result, the digital-nomad lifestyle is more easily experienced by individuals from countries who participate in visa waiver schemes. People who struggle to get visas by virtue of where they were born will find the digital-nomad lifestyle difficult to access.

There are opportunities for locals to set up businesses that support digital-nomad communities. The most obvious are coffee shops or co-working spaces. Digital nomads are always looking for places with a good internet connection and refreshments, somewhere that they can sit for hours and get their work done before going to the beach or diving or doing any of the other many things they do with their day. Establishing a co-working space can have digital nomads come flocking into an area. Many co-working spaces have expanded beyond co-working to also include co-living, setting up spaces where people don't just work but live and build a community together. KoHub, in Koh Lanta, Thailand, for example, offers private accommodation, two meals a day, and access to the co-working space for around $600 per month. The drawback to such co-living spaces, however, is that they can diminish the economic benefits of digital nomads for the wider community by centralising all of the income into one place.

What an influx of digital nomads can do is help to create and support a local tech scene. A co-working space might be set up to support a digital-nomad community, but the space is also available to locals, who often take advantage of it. The digital nomads bring outside knowledge and experience into the area which locals can benefit from. Co-working spaces often have active communities around them where people do demos, give presentations, and share knowledge. Digital nomads can help to keep a fresh influx of knowledge that locals can take advantage of.

Despite certain benefits, over-reliance on any form of tourism, whether it is mass tourism, supporting backpackers or digital nomads, can leave a host community vulnerable to future instabilities. Following the 2002 bombing in Bali, the number of foreign tourists to the island decreased by 57% and within a short period of time 2.7 million workers were left unemployed. As numbers started to increase in the following years, they were set back again in 2005 when another suicide bomb attack killed twenty-three people. Tunisia, formerly a popular destination for British tourists, has seen a drastic decrease in tourism since the 2015 terrorist attack in the town of Sousse, and in Egypt between 2010 and 2014 there has been a decrease in British visitors of 18.5% due to political unrest. In the United States, following the election of Donald Trump and policies like his travel ban, experts predict a major slump in the tourism industry. Tourism industries do tend to recover over time but during periods of high political instability millions of people can be left unemployed, businesses are destroyed, and whole industries disappear.

While you might see a backpacker reading a well-thumbed copy of Paul Coelho's *The Alchemist* or Alex Garland's *The Beach*, digital nomads have their own literary touchstones. Unlike the backpacker, whose literature is usually fiction associated with dreaming, wanderlust, and travel, you're more likely to see a digital nomad reading a book that falls into the self-help category. A backpacker is a dreamer, a digital nomad is obsessed with lifestyle. They are usually books about how to achieve your dreams, be an entrepreneur, or run your own business, books with titles like *Dream Save Do*, *The $100 Startup*, *Crush it! Why Now is the Time to Cash in on Your Passion*, *Re-work*, *The Laptop Millionaire*, and *Virtual Freedom*.

But the gospel for many digital nomads is *The 4-Hour Work Week: Escape the 9-5, Live Anywhere and Join the New Rich* (4HWW). This is the original digital-nomad book, promising to teach you how to achieve the millionaire lifestyle without being a millionaire, which means giving up your day job, using the internet to make money, and going somewhere where your money goes further. The book's author, Tim Ferris, has created a method that he calls "lifestyle design." At the heart of lifestyle design is the Pareto Principle, which states that roughly 80% of effects come from 20% of causes. Ferris applies this to life and work, saying that about 80% of what you do is unimportant, so you should focus on the 20% of causes that really have an impact. By focusing on this 20% you can jettison all of the crap in your life and design the lifestyle that you want. This is complemented by Parkinson's Law, which says that any work expands to fill the time allocated to it. Ferris advocates shortening the time available for work to just enough for the important 20% of tasks.

There are other tactics that makes Ferris' lifestyle design so appealing. He promises to use the "art and science" of lifestyle design to show you how to enter the world of the New Rich. "From leveraging currency differences to outsourcing your life and disappearing, I'll show you how a small underground uses economic sleight-of-hand to do what most consider impossible." This economic sleight-of-hand lets you live whatever lifestyle you desire, without having to break the bank to do it. This means focusing less on the end goal of amassing a fortune and being rich, and more on making enough money now to do the types of things that millionaires do. Most of these involve having adventures in different parts of the world, from chartering private planes, to skiing, to renting private villas. The basic principles involve outsourcing and automating as much of your work as possible, so you make money while having fun and travelling around the world.

It's easy to see why Ferris' book is so appealing, especially to people who are in jobs they don't enjoy or feel trapped by lives they feel have been forced upon them. Ferris promises to free you from that, to free you from the illusion of working your whole life to save for retirement, to empower you to make decisions now rather than putting them off until later, to teach you how to achieve more while doing less, to focus on your strengths not your weaknesses, to assure you that money alone is not the answer to your problems, to embrace positive stress to help you grow. "Rather than hating reality," he says, "I'll show you how to bend it to your will."

In between an overabundance of jargon and acronyms – new rich (NR), lifestyle design (LD), deferrers (D), freedom multiplier, adult-onset ADD (Adventure Deficit Disorder), DEAL, dreamlining, fear-setting–4HWW is filled with aspirational stories from Ferris' own globetrotting lifestyle and other lifestyle designers who have given up their day jobs and embraced Ferris' alternative ways of living. It is over four hundred pages long and filled with practical techniques and tools for improving your lifestyle, including ways to read faster, process more information, automate your work, delegate better, create products, and eliminate tasks.

What makes Ferris a guru in the digital-nomad world is that he defines the new rich not by their wealth but by their mobility. It is a group made up by anyone who uses the internet to work remotely, whether you are a freelancer, contractor, entrepreneur, business-owner, or employee. If you have $40,000 per year in the USA you can do a lot less with it than you can in Central or South America, or South-East Asia. An increase in the value of your money increases your choices which increases your power. Power, in this case, is of course equated with the power to make choices within the market.

The concept of geoarbitrage is at the heart of the 4HWW. Ferris defines this as: "to exploit global pricing and currency differences for profit or lifestyle purposes." Or, from an article on his blog, "Fun things happen when you earn dollars, live on pesos, and compensate in rupees." It's without a doubt one of the things that makes a digital-nomad lifestyle attractive, and was the reason I spent at least some of my time in countries where my money went a whole lot farther. Who doesn't want to do more with their money? Who doesn't want to be able to have experiences that are normally out of reach because when you are in your home country you simply can't afford them?

Digital nomads are generally well-intentioned: they want to travel and have a good quality of life. When I initially set out on my own travels I can't say that I was intentionally thinking about exploiting poorer countries. However, it is the case that I, and other digital nomads, benefit from and exploit poor economic conditions. Our willingness to take advantage of economic discrepancies and cheap local products and labour has more than a whiff of colonialism, and critics of the digital-nomad movement have no problem dubbing it "neo-colonialist." This is particularly the case when digital nomads create their own parallel communities, enjoying things that are cheap and exotic while giving little back to the host country other than spending their money there.

I recall an incident when I spent six weeks in Belize: D and I had gone for dinner at the house of some local dive instructors who we had been diving regularly with. The dive instructor lived with his family in a wooden blue house, raised by stilts from the dangers of hurricanes. When we arrived there was a queue of men outside, waiting to get their hair cut by the dive instructor's brother, who had set up an ad hoc hairdressing salon in his garden. We met the family, ate lobster, drank beer. I had taken my sunglasses off my head and set them on the table. One of the local guys picked them up and turned them over in his hand. I saw his gaze pass over the Ralph Lauren label and then a look pass between him and his friend, a set of slightly raised eyebrows and a smirk at my wealth. I didn't say anything or do anything, but inwardly I cringed. I wanted to jump up and say "I bought them in TJ Maxx!

They only cost $20! I don't have *that* much money,"
but I knew how much worse that would sound. That
no matter what the reality of my life at home was,
what I earned in a week was probably more than what
they earned in a month, and that here I was, living
the life of a rich gringo, enjoying the good weather
and the low cost of living, while they had struggled
with poor economic conditions. I felt shame, and
then shame for feeling that way, became tongue-tied
and embarrassed. But they hadn't noticed that I had
noticed, and I said nothing. And when we left we gave
them money to pay for the food and the beer. It was a
reminder of how privileged we are, that most of the
people that we met on our travels would never have
the same opportunities.

The 4HWW isn't without its insights. Ferris claims
that "the common sense rules of the 'real world' are
a fragile collection of socially reinforced illusions."
This isn't a new idea; it has taken particular force
throughout postmodernity although in the West it can
be traced back to Friedrich Nietzsche who frequently
wrote about the illusory nature of reality. "Truths," he
wrote, "are illusions about which one has forgotten
that is what they are." I can't disagree with Ferris
that our current social structures are illusory. But
lifestyle design doesn't destroy any realities or bring
to light any illusions beyond the claim that you work
in a boring job, eking out a comfortable existence,
getting by because you feel like you have to. At its heart
remains the more enduring illusions of individual
liberty and choice. It may feel like a shaking-off of
the drudgery of an old way of life for something new
and exciting, but really it's just the culmination of

forces that started to take hold in the 1970s. The Ferris archetype is a globetrotting narcissist who is focused on excitement and adventure, consuming as they go, and looking for places where they can exploit poor economic conditions. Lifestyle design is just another method for creating an individual who is cut out for riding the waves of the market.

Does lifestyle design actually result in an enduring and meaningful change to one's life? There are plenty of indications that it is a palliative, rather than a cure. Ferris writes that "Lifestyle Design is thus not interested in creating an excess of idle time, which is poisonous, but the positive use of free time, defined simply as doing what you want as opposed to what you feel obligated to do." It's this damnation of idle time that reveals one of its flaws. More than anything else it is about distraction; it creates a life of excitement and adventure which is an end in itself, concealing the emptiness, loneliness, and meaninglessness of a life in which your power to act is only the same as your power to act within the market. This doesn't mean that travel and embracing new experiences aren't worthwhile or life-changing, but they only treat symptoms. If you are unhappy and unfulfilled in your job, giving it up to travel won't solve that problem, it can even compound it as you are forced to take on less meaningful work in order to embark on this search for meaning in the rest of your life. Ferris' method espouses money and experiences, but what adds depth to our everyday lives is connections and people.

The other troubling aspect to lifestyle design is that its basic attitude is that work is anything that you don't want to be doing. Ferris' view of work is much more associated with toil and chores. He makes the assumption that anyone who is getting paid to do anything for forty hours per week is doing something that they don't enjoy. However, work can be meaningful in itself and there are plenty of people whose work gives meaning to their lives. Therefore, lifestyle design can only be targeted to those who don't find meaning in their work, who are perhaps pursuing careers for extrinsic motivations such as wealth rather than intrinsic motivations such as personal fulfilment. It's no surprise then that a study of digital nomads in Bali found that 42% of the digital nomads studied were professionals in their thirties who had left behind corporate careers often in finance and consulting.

Ferris also ignores the fact that in order for society to function, very few people can follow his lifestyle. This makes lifestyle design accessible to a limited number of people, those with the privilege of location-independence. If everyone was jetting all over the world, we wouldn't have any doctors, nurses, or teachers, no one to build houses and work in shops, no one to look after children at nursery, no one to keep the power grid running, no one to fix roads or put out fires. The very lifestyle that he designs is reliant on pilots and hotel receptionists, people who run hire companies and people who clean the airport, train conductors, waiters and waitresses, and factory workers who produce aviation fuel. His escape is predicated on the work, whether that be drudge work or meaningful work, of other people.

The digital-nomad lifestyle, while immensely attractive and potentially fulfilling, not least in the short term, is not without its own challenges. One is that the type of work that is available to digital nomads, or the type of work that they are able to do, is not that fulfilling. They often take on smaller clients and don't have the stability required to tackle big projects or to participate in extended periods of learning. My own experience reflects this. I did have a big, difficult project while I was on the road but I lacked the stability I needed to fully engage and immerse myself in it. There's something to be said for getting up day-to-day and doing the same thing. It may not provide you with the smorgasbord of new experiences available to the digital nomad, but it provides the necessary conditions for deep thinking and rumination on difficult problems. Being constantly on the road can lead to feelings of under-achievement and wasted time.

Location-independence itself is really only an illusion. While in principle you can work anywhere, in reality you can only work in places with a good internet connection, which places huge restrictions on where you can go. But internet isn't the only factor. No one can actually work on a laptop from a hammock for any length of time. People who work at a computer need an ergonomically sound space, otherwise you expose yourself to all of the health problems that come from working at a laptop with bad posture over an extended period of time. These include headaches, back pain, muscle and joint problems, Repetitive Strain Injury (RSI) and Carpal Tunnel Syndrome. The only away to avoid these strains and injuries is to set up an ergonomic workstation with your equipment positioned correctly. Life on the road isn't conducive to good ergonomics, and digital nomads who care about their health are tethered to a series of co-working spaces that cater to their needs.

Co-working spaces that cater for digital nomads are relatively homogenous. They may be in different locations, but the spaces invariably offer the same thing. They are usually spaces which market themselves as "inspiring," places you can go to "be creative," and to "meet other entrepreneurs." Of necessity, they offer facilities that you would find in any office space: desks, chairs, meeting rooms, high-speed internet, a communal kitchen, photocopying and printing, stationary like flip charts and whiteboards. Most offer free coffee. Some have hipsterish twists that are supposed to set them apart from traditional offices: a free juice or smoothie bar, ping pong tables, soft areas with bean bags, board games, free meals, organised trips and excursions.

And then there is the fact that space itself is more complex. You may be location-independent in a physical sense, but you aren't location independent in a virtual sense. You are tied to all of those virtual spaces which are essential to your work: your email, Skype, and messaging service, project-management tools like Basecamp or Trello, chatrooms like Slack or IRC. These things do not change, wherever you go. Any place that you are in is multilayered, the exotic mingled with the familiar mingled with an unchanging virtual, a concoction that prevents you from ever experiencing that true sense of authenticity and otherness that you initially set out to find.

Another side of being a digital nomad which gives lie to the sun-drenched Instagram pictures is the sense of loneliness that comes from a life on the road. Like any remote worker, digital nomads spend a considerable portion of their day online, focused on their own work or engaging with people who are in entirely different locations. Unlike remote workers who stay in one place, they don't have the opportunity to cultivate the deep ties and relationships that come from sharing a day-to-day life with different people and which form the basis for enduring and meaningful friendships. Digital nomads meet a lot of people on the road and create a lot of connections. They have a broad network of friends and peers. It is wide but it is shallow. Despite being hyper-connected you lose that sense of deep connection with a few other people. There is plenty of research that emphasises that in-person time is necessary for an enduring relationship. While they are on the road, friends and family back home are celebrating the big things like births and marriages, but also the small things like birthdays and graduations, seeing children grow up and being part of their lives, seeing the same people for coffee every week. In the time you are away, people change and grow, and you yourself return a different person. On a post on Medium, programmer Charlie Guo writes: "while I've been gone, my old friends have changed. I've missed my fair share of celebrating, commiserating, and reminiscing. Facebook has become a window into birthdays, graduations, and housewarmings; so close and yet so far. I'd love to be there in person, making inside jokes and new memories, but I've traded that for passport stamps and culture shocks."

There are other challenges that digital nomads rarely talk about, including; low job satisfaction, slow career development, high administration overhead, overcoming negative perceptions of digital nomads, just the illusion of independence, and a lack of any traditional employment support such as peer support networks, collegiality, feedback, and mentorship. This makes it hard to sustain a digital nomad lifestyle for an extended period of time. This is compounded by the fact that many nomads only earn enough to sustain themselves and aren't saving money for their future.

Let's return again to Pieter Levels, the plucky young entrepreneur who was profiled in WIRED in 2014. In 2016, there was a new article about Levels, this time in QZ. In the article, Levels recalls the lack of connection he felt when looking out the window of his apartment in Medellin, Columbia. Despite being the epitome of a successful digital nomad, life was missing something. "I started feeling lost. I started asking, 'Who am I?' A large part of [your identity] is your environment. When you're moving around from place to place, and you aren't making long-term friends, you lose a big part of your identity. I'm a pretty strong and stable person, but I wasn't prepared for that." Despite planning to travel again in the future, Levels acknowledges that he needed to put down roots.

One of the things I'm most thankful for that I got from my year as a digital nomad was that I learned to scuba dive. D and I did our initial certification in Thailand in late January, and by the time we did our final dive in Mexico in July we had logged more than a hundred dives each. This had been a lifelong ambition of mine and it lived up to every expectation. From diving with whale sharks in Thailand, to exploring wrecks off the coasts of New Zealand, to rescuing D during a night dive in Cozumel, to swimming with sharks, to nitrogen narcosis at the Blue Hole, to the terror and otherworldliness of the Angelita cenote in the Mexican jungle, diving during my travels as a digital nomad was one of my life's greatest experiences. It would be churlish of me to suggest that this was enabled by anything other than the fact that I could take my laptop and could make money on the road, working one day and diving into the blue on the next. There is no other way that we could have afforded to dive so much and in such varied places in such a short space of time. It is, undoubtedly, a designed lifestyle that Tim Ferris would be proud of.

But, though they were all fun, it wasn't the adventure, the adrenaline, and the excitement that drove me to dive again and again. Diving brought, and continues to bring, something increasingly absent from my life. Though I can recall numerous underwater moments this one stands out in particular. We were diving in Cozumel in Mexico – it was getting to the end of the dive and most of the group had already ascended due to being low on air. Myself, D, and our dive guide, Joe, still had half-full tanks so we continued to explore. Joe motioned to us from the other side of the reef. D and I swam over to him and descended until we hovered just a few inches off the sandy ocean floor. Under the rock, in a shallow cave, was a nurse shark that looked to be sleeping. It was about the same size as me, rested on the sand, its mouth opening and closing so that water rushed through its gills. It barely moved, just coiling its tail once or twice. The three of us hovered there with the nurse shark, watching it, silent, suspended in the water, our breathing controlled and shallow so that our air would last as long as possible. Poised there, I lost track of time, lost track of myself, I was just my breath and a sleeping shark, the rest of the world gone, irrelevant.

We pushed our air for as long as we could, and when we eventually were forced to ascend, we did so with a sense of tranquillity, as if for a short time we had exited one world and been part of another. It may sound extreme, but the only way I could disconnect fully was to throw on some diving equipment and descend into the sea. There is total disconnection, not just because I cannot access the internet and no one can access me, but because I am forced to focus only on the subtle and slow movements of my own body as I glide weightless, suspended in water, and on the wildlife on which I am an interloper. When I am underwater I feel like I am no longer implicated in a mass-communication network, another node, another point of interaction, a being that is distracted and is connected. I feel like nothing, just a body, removed and forgotten and dissipated and peaceful. By focusing my mind on just one thing, I was able to think more clearly and feel more peaceful than I had done in a long time.

That sense of disconnection is something many people are lacking. That we crave it as a society is illustrated by the rapid growth of the mindfulness industry which generated around $1 billion in revenue in 2015. People want to disconnect from the fast-paced stream of technology and free markets, to find ways to better be at one within themselves and at peace. This is why Buddhist-inspired mindfulness techniques have become so popular not just at the individual level but at the corporate level, with companies providing mindfulness training for their employees. This Western approach to mindfulness uses Buddhist techniques without the ethos; rather than freeing one from greed, selfishness, and ill-will, they are just another tool for self-betterment. Whether it's mindfulness or scuba diving, these solutions operate at the level of the individual. They are fixes for you and I, but don't address the societal problems that have led to us feeling connected but isolated. They are pragmatic steps, short-term fixes that relieve anxiety and pain, and are intended to achieve clear, understandable goals. Mindfulness gives people the tools to address toxic and high-pressure situations so that those situations themselves don't need to be remedied. Stress and anxiety are framed as individual problems, rather than being symptoms of a wider malady within society.

The digital-nomad lifestyle is one that is attractive and aspirational. It's not hard to see why so many, like me, take to the road in search of a better quality of life, one that is more engaged, less stressful, and more fulfilling. We are thrown into a society to which we are constantly expected to adapt. Not a day passes when the political, economic, and technological landscape doesn't feel different and uncertain. We have inherited a world that is more precarious, one where we are expected to be flexible and be constantly adapting. When we take to the road we take control of our own lives, go where we want, make our own decisions. It is a refusal of the failed promises of the previous generation, of stability and constant upwards growth; it is a way of adapting that gives us the feeling of hope, of promise, and of escape.

But what many digital nomads find is that life is made meaningful through a different type of connection than that created by cables and bytes. The thing that often galvanises them is not the lifestyle that they have created for themselves, but that they are given an identity by being one of a tribe, that they take strength in a collective, that they can be more than just themselves by being among others with similar goals and aims. Beneath our obsession with the individual is the truth that we are more when we are together, and that if we harness that collective energy we can do more than we do alone. It's sad though that we continue to ignore that potential and instead continue to design lifestyles that are just for ourselves.

LIFE'S GREAT BEAUTY

In Paolo Sorrentino's 2013 film *The Great Beauty*, Jep Gambardella is an ageing playboy writer. His one novel, *The Human Apparatus,* was published in his twenties to great acclaim, but now at sixty-five he has failed to write another. In the interim forty years, Jep has lived a hedonistic lifestyle in Rome, going to sexy parties, sleeping with beautiful women, making money by writing articles and doing interviews with second-rate artists, and living off the money and prestige afforded by his first novel. His life has been one of distraction. Rather than live up to his promise, he has stayed on the surface, embracing a life of enjoyment.

At the movie's end, a wizened nun who has fallen into his social circle asks, "Why did you never write another book?"

Jep responds, "I was looking for the great beauty, but I didn't find it."

The nun responds, "Do you know why I only eat roots?"

"No, why?"

"Because roots are important."

Roots go beneath the surface of things, they are deep and searching and penetrate into darkness. Jep was never able to find life's great beauty in his distractions. Beauty, great beauty that is worth writing a book about, cannot be found on the surface of things, in life's distractions. It can only be found in the depths, by looking deeply into the world and taking all its darkness with all of its beauty.

The Great Beauty is about more than the life of one writer. It is a story about how we all start with great potential, a great capacity for insight and depth, but when we allow ourselves to be permanently distracted we miss out on what is truly beautiful about life. The internet provides more opportunity for distraction than we have ever had before. When we engage with it thoughtlessly and uncritically we are in danger of missing out on so much. We distract ourselves from personal problems that never go away; around us the world is changing and we are changing with it.

The internet is a playground of distraction. No one needs to sell all of their belongings and go on the road to distract themselves. We can all pick up our phones, play Candy Crush, glance over Twitter, send a text message, see what's happening on Facebook, we can watch endless television series on Netflix, or pass the time chatting with strangers all over the world. We live in a world that has endless novelty and endless distraction, giving us constant opportunity to avoid problems that seem insoluble, to become forgetful, to favour novelty over meaning.

When we think about the internet, we usually talk about what we have gained, the things that it has added to our lives: access to knowledge and information, greater opportunity to connect to other people, access to a wider range of goods, convenience, novelty, speed, and work opportunities. The sheer amount of data helps us to make decisions, helps us to craft our lives and feel like we are in control. It's not a contradiction to acknowledge that while all of these things may make our lives better, there is still much that we have lost. In the transition to a hyperconnected self, there are aspects of ourselves that feel like a distant memory.

I miss solitude. Real, disconnected solitude. The type of solitude which, as Jep Gambardella learns in *The Great Beauty*, is needed to write a great novel, to engage in deep thought. Going out and leaving my phone at home isn't enough. There's always a message waiting somewhere, always the potential of a connection. Before the internet was in my pocket, I recall daydreaming through a train journey, while waiting for a bus, lost in my thoughts while walking down the street. Now it's hard to imagine ever truly being alone.

Writing and researching this book has been a practical exercise in finding the solitude I crave. It has been a struggle against distraction, a search for solitude, a rumination on loneliness. I would always rather be playing a game on my iPhone, chatting online, working, doing anything other than sitting down to think and write. I want to be distracted and there is so much joyful opportunity for it. I've found my writing and research to be scattered, pulled in different directions, packet-switched, working on one idea for a short period, drawing a connection to something else, writing a different section, then back, unsure where I started. This book has been written in a bricolage fashion, pulling together ideas and concepts in an attempt to create something coherent out of a mass of information. Whether it has been successful or not is up to you. But as a writer I have found myself scattered and then pulled together, scattered and then pulled together, writing in a mixture of fragments and longer essays, skipping between sections and hopping between sentences. The most productive I have felt while writing has been when I have been able to focus in on long sections, forcefully bracketing off the constant and flickering barrage of ideas that are always there, always on the periphery, beckoning me down another rabbit hole.

Those moments of solitude that I've been able to find, whether that's sat on the beach or staring out the window of my office, have created a space for critical reflection. Despite living in a world in which everyone has an opinion and is willing to share it, there seems to be very little space for deep thought. A lot of what is written on the internet is reactive. An article is written in response to someone else's article. Half of news stories, even in mainstream media, just seem to be reports of what people say on Twitter, the other half it's hard to tell if they are true or false. We distract ourselves by consuming information; in this endless consumption are we losing touch with the capacity that we all have for critically reflecting upon the world around us?

In my generation, we have transitioned to a world abundant with information. Information and knowledge are available to everyone. It is a triumph of the commons. Where once you would have had to go to university to gain knowledge, now anyone can access any piece of information. All you need is a search engine. But while we have gained much in terms of access, what we haven't gained alongside it are the cognitive tools to deal with this level of information. Our school curricula teach us that knowledge given from authority is objectively true, the media presents its narrative as facts, and governments expect us to uncritically accept their stories. We are presented with a mass of contradictions, opinions presented as facts coming from every different direction. What we all need to gain, and what should be taught to our children, are the critical skills to filter, deal with, and reflect upon the high level of information that we are daily bombarded with. When we get this critical distance we are able to unlock the internet's potential as an educational tool. When we learn to question the information presented to us, when we are better able to distinguish between truth and falsehood, then we have an abundance of education at our fingertips and greater scope for learning than ever before.

When we are able to take a pause, edging ourselves just slightly away from full immersion, we can not only better reflect on our being and our work, but also get a better understanding of the true, emancipatory power of the internet. The internet is not just social media and online magazines, it is not just a place for advertisers to make money or commerce to happen, it is not just a place for watching movies or looking at porn or sharing cat videos. It is a place where connections happen. These connections make it possible for national and international grassroots organisation on a scale that has never before been possible. This can be seen in the ways that people use internet tools to organise for collective action. Examples vary widely, from the use of blogs and social media in uprisings in the Middle East; to support work such as CalAid, the group set up in 2015 to support refugees in Calais; and protests like the Women's March, a worldwide march for women's rights. Collective action also takes place outside of politics, in the way that people form support groups on platforms like Facebook, and in the free software movement, where people come together to create software that is available to everyone. A shift from the language of connectivity to collectivity would transform the discourse in a way that would disrupt our everyday internet usage and help us to use it to its full potential.

Are there ways to use the internet to create a more equitable society? Can we create disruptions that don't just benefit business but that benefit everyone? Can we use the internet to improve the working conditions for everyone, whether they be people who work online or not? What do we need to do to resist the marketisation of the internet? How can we use the internet to enhance our lives without quite so much subtraction? The story of the internet is still being written and we are all its authors; we just need to sometimes take a step back, pause, and think about what we are writing.

D and I returned home and our lives took another turn. Four months after landing back in Kent I was pregnant, and we prepared for the transition that has been experienced by new parents for as long as there have been human beings. Of necessity, my life had to change. I couldn't just live the life of last-minute travel plans and long-haul flights. I needed to be at home. Morning sickness grounded me in the first trimester and other than a few last trips I spent the months before the birth of our son preparing for his arrival. When he came I just didn't want to be away anymore, not as I had been; I wanted my life with my family.

By that time, I had developed some of the most essential skills for any remote worker: the ability to get a critical distance from my work and create boundaries that keep it out of the rest of my life. It is rare for me to work past 5pm, my weekends are my own, my phone is generally free of any work tools, and I am able to bracket off work even within my thoughts. As many remote workers no doubt find, this comes from a mixture of experience and reflection upon what really matters in your life.

But, in the months running up to the birth of my son and afterwards, one of the biggest disadvantages of remote work remained: I was lonely, and more than I had ever been. I may have had online friends but I had no one around me, none of the in-person network that provides such crucial support for new mothers. I had no family nearby, no friends within a hundred miles. I made an effort to make new friends, went to antenatal groups, but without the familiarity of the internet I found it difficult to make connections.

The technology that had liberated me felt like a trap. It had enchanted me with its freedom and speed and flexibility but failed to address my human need for real connection. I had been tricked. I could live anywhere, work anywhere, but working online could never provide me with all of my needs, and I was lonely. The year that D and I spent travelling had been packed full of incredible experiences, but it was for all that still a distraction. We returned home to many of the same problems that we thought we had left behind. We had just postponed dealing with them while we distracted ourselves with shiny new things.

Shortly after the birth of our son we decided to deal with it. We once more took advantage of my location-independent status. D quit his job and we moved to the other end of the country, to a place where I had friends and connections that weren't reliant on the internet.

I'm not sure if we'll ever put down permanent roots, or whether that's even possible in the turbulent world we live in, but we are as close to settled as we have ever been. One of the best parts of living in our new home is that we are so close to the sea. I can see it from my home office, navy blue and still, blending into the horizon. It feels far away from the stormy blues of the English Channel, at which I used to stare while working in my hated admin job.

I walk on the beach often, using the titanic sea as a way to disconnect from the network, turning away from one sense of overwhelming vastness to another. Since leaving my office job I have seen, and jumped into, many different oceans, the greens of the Andaman, the depths of the north-west Pacific, the sparkles of the Caribbean. Each is distinct yet connected, a global body of water in which small changes have far-reaching effects. It makes me feel small. When I stand on the beach with the sounds of the waves crashing and dogs barking, I feel truly connected, with a depth and stillness that is lacking in the oceanic mass of the internet. I feel anchored, I feel a connection to the earth, a tiny fragment in the passing of geological time. The sea air blows away the blah, blah, blah, and I can fix my gaze on a horizon that leads nowhere. It is the pause that I so often need, a moment of presence, of keeping at bay the chattering of the many, the Sirens that call from the network, a moment to sink into peace, oblivion, and silence.

BIBLIOGRAPHY

This book draws upon research from a wide range of disciplines. Rather than break the flow of reading, I chose not to include hyperlinks or footnotes. References to all of the works cited can be found on my website at http://siobhanmckeown.com.

There were many books and articles that helped guide the ideas I've written about here. The most important are:

Ballard, J.G. *High-Rise.* Harper Perennial, 2006.

Brockling, Ulrich. *The Entrepreneurial Self.* Sage Publications Ltd, 2015.

Carr, Nicholas. *The Shallows: How the Internet is Changing the Way We Think, Read and Remember.* Atlantic Books, 2011.

Ciulla, Joanne B. *The Working Life: The Promise and Betrayal of Modern Work.* Crown Publications, 2001.

Harvey, David. *A Brief History of Neoliberalism.* Oxford University Press, 2007.

Mazzucato, Mariana. *The Entrepreneurial State: Debunking Public vs. Private Sector Myths.* Anthem Press, 2015.

Foucault, Michel. *Discipline and Punish: The Birth of the Prison.* Penguin, 1991.

– – – . *The Birth of Biopolitics: Lectures at the Collège de France, 1978-1979.* Palgrave MacMillan, 2010.

Smail, David. *The Nature of Unhappiness.* Robinson, 2001.

Thompson, E.P. "Time, Work-Discipline, and Industrial Capitalism." *Past & Present:* No. 38 (Dec. 1967), 56–97.

Turkle, Sherry. *Reclaiming Conversation: The Power of Talk in the Digital Age.* Penguin Books, 2016.

ACKNOWLEDGEMENTS

First and foremost, my thanks to Darren Ambrose, my husband and partner, who has read multiple versions of this book, providing ever-insightful feedback, critical insight, and support through the writing process. He never lets me take myself too seriously while encouraging me to be the best I can. But, more importantly, he lives this life with me. Much of what I have written is the result of our own struggles and conflicts, both with each other and within ourselves. This book is one story of our life together, and our discussions and reflections upon that life. I am grateful to him for his love, his constancy throughout our relationship, and for letting me share so much of that relationship here.

This book is a record of my work online, which has been made possible by far too many people to mention. However, there are some who have played a pivotal role in my working life, and those who have inspired me in various ways, either by their actions or through our long conversations and agonising about working online. My thanks to Hampton Catlin, Mason James, James Farmer, Boone Gorges, Michael Pick, Zé Fontainhas, Samuel Sidler, Matt Mullenweg, Hanni Ross, Mike Schroder, Brian Krogsgard, Andrea Middleton, and Krista Stevens. At various points, your influence on my work has been great, and without you my life, and this book, would be quite different. I also want to thank everyone at Human Made for making me reflect daily on the nature of what we do, and particularly Tom Willmot for letting me put those reflections into action.

My thanks to Milan Jaros for his early insistence that philosophy is not just in books but that we find it in the world, by analysing those things around us. And, of course, my thanks to Tariq Goddard, my editor and publisher at Repeater, for publishing this book which has been simmering in my head for a number of years, for being a thoughtful interlocutor, and for encouraging me to embrace my own optimism.

In 2004, via my blog, I made my most enduring online friendship. I was both flattered and excited when Mark Fisher, aka k-punk, linked to my blog, and from that first connection we quickly became friends. In the interim years we spent many days together talking and arguing, hanging out, laughing, reading books, going for walks in the Suffolk countryside, watching reality television, and enjoying (ish) the occasional caravan holiday. Mark read the original proposal for this book and encouraged me to send it to Repeater.

In January 2017, when I was completing the final version of this book, Mark took his own life. There are no words to say how devastated I felt and how I still feel. It brought home the realities of depression and mental illness, issues that are touched upon in this book. Mark was someone who I could always rely on for a sense of hope when things felt hopeless. He lived the possibility that things could be different, that we have the power to change this world if we try, and that thought, critical thought, is one of the most essential weapons in our arsenal. If I could, I would remind him of the sense of hope that he has given me, I would thank him for being such a positive influence on my life, and for encouraging me to trust my own ideas and critically engage with not just books but the world as I find it. It saddens me that he will never read this book that is so infused with our years of friendship, but I hope that he would have found it worth reading, and at times worth arguing with. I don't think this book would have been the same, or even possible, without him; it is written very much in his spirit.

Repeater Books

is dedicated to the creation of a new reality. The landscape of twenty-first-century arts and letters is faded and inert, riven by fashionable cynicism, egotistical self-reference and a nostalgia for the recent past. Repeater intends to add its voice to those movements that wish to enter history and assert control over its currents, gathering together scattered and isolated voices with those who have already called for an escape from Capitalist Realism. Our desire is to publish in every sphere and genre, combining vigorous dissent and a pragmatic willingness to succeed where messianic abstraction and quiescent co-option have stalled: abstention is not an option: we are alive and we don't agree.

Repeater